PCR Sequencing Protocols

METHODS IN MOLECULAR BIOLOGY™

John M. Walker, SERIES EDITOR

METHODS IN MOLECULAR BIOLOGY™

PCR Sequencing
Protocols

Edited by

Ralph Rapley

Coventry University, Coventry, UK

Humana Press ✳ **Totowa, New Jersey**

© 1996 Humana Press Inc.
999 Riverview Drive, Suite 208
Totowa, New Jersey 07512

For additional copies, pricing for bulk purchases, and/or information about other Humana titles, contact Humana at the above address or at any of the following numbers: Tel.: 201-256-1699; Fax: 201-256-8341; E-mail: humana@interramp.com

All authored papers, comments, opinions, conclusions, or recommendations are those of the author(s), and do not necessarily reflect the views of the publisher.

This publication is printed on acid-free paper. ∞
ANSI Z39.48-1984 (American Standards Institute) Permanence of Paper for Printed Library Materials.

Cover illustration: Fig. 1 from Chapter 8, "Direct DNA Sequencing of PCR Products Using Magnetic Beads," by Joakim Lundeberg, Bertil Pettersson, and Mathias Uhlén.

Photocopy Authorization Policy:

Printed in the United States of America. 10 9 8 7 6 5 4 3 2 1

Library of Congress Cataloging-in-Publication Data

Main entry under title:

Methods in molecular biology™.

PCR sequencing protocols / edited by Ralph Rapley.
        p. cm. — (Methods in molecular biology ; v. 65)
    Includes index.
    ISBN 0-89603-344-9 (alk. paper)
    1. Polymerase chain reaction—Laboratory manuals. I. Rapley,
Ralph. II. Series: Methods in molecular biology (Totowa, NJ) ;
v. 65.
QP606.D46P365 1996
574.87'3283—dc20   96-27724
                                                  CIP

Preface

Advances in bioscience research usually arise as a result of the continuing refinement of existing technologies. However, there are a number of occasions where newly developed methodologies have a profound effect on nearly all areas of research. Frequently these are techniques that are elegantly simple in concept and require minimal technical manipulation. Two of these revolutionary techniques are the focus of *PCR Sequencing Protocols.* The first such technique is enzymatic chain termination sequencing developed by Sanger and his co-workers in Cambridge and reported in 1977. This essentially brought the possibility of deriving nucleotide sequence information in a very short time scale and has been widely accepted in many laboratories as a routine molecular biological research tool. Furthermore, it has not only led to the sequencing of many genes and gene fragments, but has also allowed the technical means of sequencing the human genome.

The second technique that has found widespread acceptance in basic applied research and many routine applications is the polymerase chain reaction. This technique, first reported in 1985 by Mullis and his colleagues, provides the means to amplify nucleic acid sequence, which immediately proved invaluable in nearly all fields of biological laboratory research. Here, as with enzymatic DNA sequencing, is a very simple concept that relies on minimal information to prepare short oligonucleotide primers that direct the synthesis of a specified fragment of DNA in the presence of a thermostable DNA polymerase.

The fact that these two methods lend themselves well to automation has contributed to their widespread acceptance. Together these two techniques provide the most rapid way of generating nucleic acid sequence information and consequently numerous variations and refinements of methods have been developed. The aim of *PCR Sequencing Protocols* is to bring many of the accepted protocols for PCR sequencing together. Essentially the volume focuses on three broad areas of PCR sequencing that are not mutually exclusive, but overlap to a certain extent. Many of the direct PCR sequencing protocols deal with the problem of the rapid reassociation of amplified complementary strands by modifying the sequencing reactions. This is detailed in the earlier chapters, in addition to PCR purification, primer labeling, and preparation of sequencing gels. This leads to more complex methods of primer manipulation to enable affinity purification or transcript production providing homogeneous single strands. The latter chapters deal with more conventional means of cloning PCR products into vectors and their subsequent sequencing. More extensive protocols for conventional cloning and sequencing may be found in *DNA Sequencing Protocols* (*Methods in Molecular Biology,* vol. 23).

PCR Sequencing Protocols has been prepared in the same format as other volumes in this series in that it not only provides easy and clear instructions on the protocols presented, but also important practical points or notes on the individual intricacies of the methods. These are frequently small, but important, details that lead to a successful outcome with a particular protocol. I would like to thank all those involved in the preparation of this volume, my colleagues for helpful suggestions and Professor John Walker, the series editor, for all his help and encouragement.

Ralph Rapley

Contents

Contributors

BENTLEY A. ATCHISON • *Victorian Institute of Forensic Pathology, Monash University, South Melbourne, Australia*

BARBARA BACHMANN • *Deutsches Primatenzentrum, Abteilung fur Virologie and Immunolgie, Gottingen, Germany*

PETER B. BECKER • *Gene Expression Programme, European Molecular Biology Laboratory, Heidelberg, Germany*

NEIL BREWIS • *Department of Pathology, School of Medicine, University of California, San Diego, La Jolla, CA*

FRANK C. BROSIUS III • *Departments of Internal Medicine and Nephrology, University of Michigan Medical School, Ann Arbor, MI*

XINAN CAO • *Ann Arbor Veterans Administration Hospital, Ann Arbor, MI*

JEAN-LAURENT CASANOVA • *Developpement Normal et Pathologique, INSERM U132, Hopital Necker, Paris, France*

C. THOMAS CASKEY • *Merck Research Labs, Merck and Co. Inc., West Point, PA*

ALISON COFFEY • *The Sanger Centre, Cambridge, UK*

SUSAN E. DANIELS • *Wellcome Trust Centre for Human Genetics, Oxford, UK*

ALBERT B. DEISSEROTH • *Department of Neuro-oncology and Department of Haematology, MD Anderson Cancer Center, The University of Texas, Houston, TX*

ANDREA M. DOUGLAS • *Bone Marrow Research Laboratories, Royal Melbourne Hospital, Melbourne, Australia*

IAN DUNHAM • *The Sanger Centre, Cambridge, UK*

RUPERT EGENSPERGER • *Laboratory of Molecular Neuropathology, Ludwig-Maximilians-University, Munich, Germany*

MORTIMER M. ELKIND • *Deprtment of Radiological Health Sciences, Colorado State University, Fort Collins, CO*

MANUEL B. GRAEBER • *Laboratory of Molecular Neuropathology, Ludwig-Maximilians-University, Munich, Germany*

BARBARA ANNE HALES • *Division of Biological Sciences, School of Natural Sciences, Coventry University, Coventry, UK*

HOLLY A. HAMMOND • *Merck Research Labs, Merck and Co. Inc., West Point, PA*

LAWRENCE B. HOLZMAN • *Department of Internal Medicine, University of Michigan Medical School, Ann Arbor, MI*

GERHARD HUNSMANN • *Deutsches Primatenzentrum, Abteilung fur Virologie and Immunolgie, Gottingen, Germany*

GABOR L. IGLOI • *Institut fur Biologie, University of Freiberg, Germany*

BERNHARD KALTENBOECK • *Department of Pathobiology, College of Veterinary Medicine, Auburn University, AL*

SIEGFRIED KÖSEL • *Laboratory of Molecular Neuropathology, Ludwig-Maximilians-University, Munich, Germany*

KONSTANTIN G. KOUSOULAS • *School of Veterinary Medicine, Louisiana State University, Baton Rouge, LA*

TAMMY LIND • *Department of Biochemistry and Molecular Biology, Mayo Clinic/Foundation, Rochester, MN*

JINGMEI LIU • *Life Sciences Division, Los Alamos National Laboratory, Los Alamos, NM*

CHRISTOPH B. LÜCKING • *Laboratory of Molecular Neuropathology, Ludwig-Maximilians-University, Munich, Germany*

WOLFGANG LÜKE • *Deutsches Primatenzentrum, Abteilung fur Virologie and Immunolgie, Gottingen, Germany*

JOAKIM LUNDEBERG • *Department of Biochemistry and Biotechnology, Royal Institute of Technology, Stockholm, Sweden*

IVOR J. MASON • *Division of Anatomy and Cell Biology, UMDS Guy's Hospital, London, UK*

GEORGES-RAOUL MAZARS • *Ludwig Institute for Cancer Research, University College, London, UK*

MARCIA A. MCALEER • *Department of Molecular Biology, Yamanouchi Research Institute, Oxford, UK*

MARY I. COOLBAUGH MURPHY • *University of Texas Health Science Center, Texas Medical Center, Houston, TX*

HARALD PETRY • *Deutsches Primatenzentrum, Abteilung fur Virologie and Immunolgie, Gottingen, Germany*

BERTIL PETTERSSON • *Department of Biochemistry and Biotechnology, Royal Institute of Technology, Stockholm, Sweden*

ERAN PICHERSKY • *Department of Biology, University of Michigan, Ann Arbor, MI*

JEAN-PIERRE QUIVY • *Gene Expression Programme, European Molecular Biology Laboratory, Heidelberg, Germany*

RALPH RAPLEY • *School of Natural Sciences, Coventry University, Coventry, UK*

ZHIYUAN SHEN • *Life Sciences Division, Los Alamos National Laboratory, Los Alamos, NM*

STEVE S. SOMMER • *Department of Biochemistry and Molecular Biology, Mayo Clinic/Foundation, Rochester, MN*

ANU SUOMALAINEN • *Department of Human Molecular Genetics, National Public Health Institute, Helsinki, Finland*

ANN-CHRISTINE SYVÄNEN • *Department of Biochemistry and Biotechnology, Royal Institute of Technology, Stockholm, Sweden*

CHARLES THEILLET • *Ludwig Institute for Cancer Research, University College, London, UK*

BIMAL D. M. THEOPHILUS • *Department of Haematology, The Birmingham Children's Hospital, Birmingham, UK*

ERIK C. THORLAND • *Department of Biochemistry and Molecular Biology, Mayo Clinic/Foundation, Rochester, MN*

MATHIAS UHLÉN • *Department of Biochemistry and Biotechnology, Royal Institute of Technology, Stockholm, Sweden*

ALISON WADE-EVANS • *Institute of Animal Health, Pirbright, Woking, Surrey, UK*

DAVID WALSH • *Department of Molecular Genetics, School of Biological Sciences, University of Auckland, New Zealand*

ROBERT L. WELLS • *Deprtment of Radiological Health Sciences, Colorado State University, Fort Collins, CO*

CRAIG WINSTANLEY • *Division of Biological Sciences, School of Natural Sciences, Coventry University, Coventry, UK*

WEI ZHANG • *Department of Neuro-oncology and Department of Haematology, MD Anderson Cancer Center, The University of Texas, Houston, TX*

1

Preparation and Analysis of DNA Sequencing Gels

Bimal D. M. Theophilus

1. Introduction

DNA sequencing involves a specific application of electrophoresis to resolve the linear single-stranded fragments produced during sequencing reactions, which differ in length by a single base pair. This necessitates using an acrylamide gel, usually at a concentration of 4–20%, of at least 40 cm in length and normally 0.4 mm thick.

Sequencing gel solutions are poured into a mold comprising two glass plates held apart by plastic spacers that run the length of the plates at their edges. A variety of methods are available for sealing the sides and bottom edge of the mold to prevent leakage of the gel mix before polymerization. Different manufacturers of sequencing gel kits often incorporate their own particular design features for achieving this. During electrophoresis, the mold supports the gel in a vertical position in the tank (Fig. 1A). One plate is shorter than the other at the top of the gel to form an upper buffer chamber into which samples are loaded.

The wells into which samples are loaded may be formed by a standard comb whose rectangular teeth form indentations in the gel. More commonly employed is the "shark's tooth" comb, which has a straight edge and a jagged edge comprising 24–48 triangular teeth (Fig. 1B). The shark's tooth comb is advantageous because there is virtually no separation between adjacent lanes of a sequence in the final autoradiograph. Also, samples may be loaded more simply with a Pipetman, rather than by using a 0–10 µL syringe, which is necessary when using conventional combs. The straight edge of the shark's tooth comb is used to form a flat, uniform surface across the top of the gel while setting. For running, the comb is reversed so that samples may be loaded into wells partitioned by the points of the teeth.

From: *Methods in Molecular Biology, Vol. 65: PCR Sequencing Protocols*
Edited by: R. Rapley Humana Press Inc., Totowa, NJ

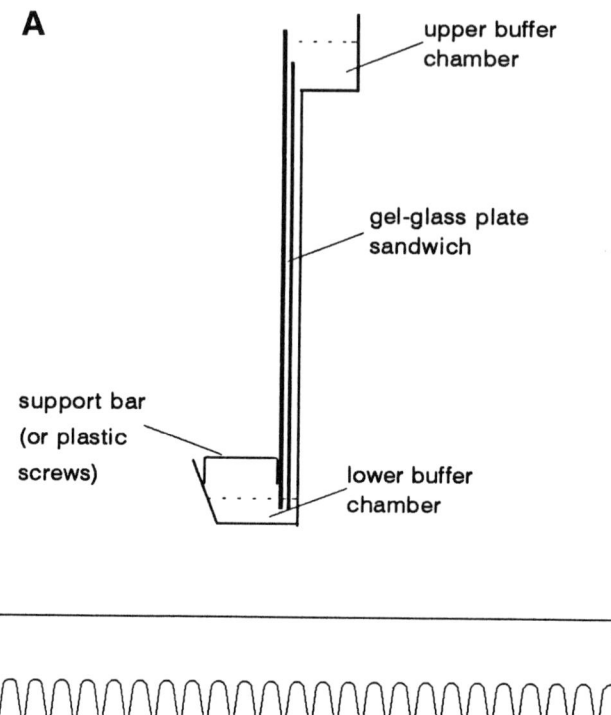

Fig. 1. (A) Sequencing gel apparatus. **(B)** Shark's tooth comb.

The radiolabeled fragments produced during the four chain-termination reactions are run on adjacent lanes of the gel. Several steps must be taken to prevent DNA from forming stable secondary structures by self-hybridization: The samples are heated to at least 70°C in the presence of the denaturant formamide before loading, the denaturant urea is incorporated into the acrylamide gel at 7M, and the gel is run at around 50°C. Despite this, artifacts owing to secondary structure may still be apparent on electrophoresis (*see* Section 3.5.).

The samples undergo electrophoresis for an appropriate length of time, which is determined by the distance between the primer used to prime the sequencing reactions and the region of sequence to be analyzed. The gel is then dried and exposed to X-ray film. A sequence "ladder" is produced from which the sequence of the DNA template can be determined by reading successive bands of increasing size in the four adjacent tracks of the gel.

2. Materials

1. Sequencing gel apparatus, comprising:
   a. Gel tank.

b. Electrical leads (usually incorporated into tank safety covers).
c. Glass plates (usually 21–38 cm wide × 40–100 cm long).
c. 2× 0.4-mm side spacers.
e. 1× 0.4-mm bottom spacer (optional).
f. 0.4-mm standard or shark's tooth comb.
g. Clamps or bulldog binder clips.
2. Gel dryer.
3. 10X TBE buffer: 108 g Tris base, 55 g boric acid, 9.3 g Na$_2$EDTA·H$_2$O, in 1 L deionized H$_2$O. (The pH should be around 8.3, without adjustment.)
4. 30% Acrylamide: 28.5 g acrylamide, 1.5 g bis-acrylamide. Make up to 100 mL with deionized H$_2$O, filter, and store at 4°C (*see* Note 1).
5. Freshly made 25% ammonium persulfate: Dissolve 0.25 g in 1 mL distilled H$_2$O.
6. *N,N,N',N'*-Tetramethylethylenediamine (TEMED).
7. Urea.
8. X-ray film and cassettes.
9. Siliconizing solution (dimethyl dichlorosilane).

3. Methods

3.1. Assembling the Gel Plates

1. Ensure all parts of the apparatus are thoroughly clean (*see* Note 2).
2. Place one glass plate horizontally on a bench "inner" side facing upward (*see* Note 3). Place the clean, dry spacers along the long edges and along the bottom edge if one is provided for this purpose.
3. Place the other glass plate on top of the spacers, so the two "inner" sides are facing each other. The bottom edges of the plates and spacers should be aligned.
4. Secure the assembly together along the sides with bulldog clips spaced approx 2 cm apart. If the bottom edge has a spacer, this should be similarly clamped. If not, seal the bottom edge with waterproof tape. The tape may also be used in combination with clips along both the sides and bottom edges for added security, especially near the bottom corners, which are particularly prone to leakage (*see* Note 4).

3.2. Pouring the Gel

1. To pour a 6% gel, combine 63 g urea, 15 mL of 10X TBE, and 30 mL of 30% acrylamide stock solution (*see* Note 5). Make up the volume to 150 mL with distilled H$_2$O This solution can be made up as a stock and stored at 4°C for several weeks (*see* Note 6).
2. Add 50 µL each of 25% ammonium persulfate and TEMED to 50 mL of the urea/TBE/acrylamide gel mix, which has been allowed to warm to room temperature. Mix by swirling. This volume is sufficient for a 21 × 50 cm plate assembly, but an additional aliquot (10–30 mL) may be required for systems that recommend sealing the bottom edge with acrylamide before pouring the main gel (*see* Note 4).
3. Without delay, take the gel mix into a 50-mL syringe, attach a needle, and inject the mix between the plates, maintaining a steady flow. During pouring, the plates

should be supported by the left hand at a 30° angle and to the side so that the corner into which the mix is injected is uppermost, whereas the diagonally opposite corner is resting on the bench. Any air bubbles that form should be removed immediately by gently raising the glass plate to lower the level of the liquid, gently tapping the plates with a finger or Pipetman, or using the comb to draw the bubble to the surface.

4. Once the gel solution has reached the top, rest the assembly at about 5° to the horizontal (for example, on a roll of sticky tape placed near the top of the assembly).

5. Insert the straight edge of the shark's tooth comb about 5 mm into the gel. If a traditional rectangular-toothed comb is used, the toothed edge should be inserted into the gel. It is also advisable to clamp the glass plates over the comb with two bulldog clips to reduce the risk of leakage across adjacent lanes during sample loading. Check the gel over the next few minutes, and top it up as necessary using the gel mix remaining in the syringe.

6. The gel should set within 1 h, but to maximize resolution, it is recommended to age the gel for at least 3 h before use. If the gel is to be left overnight, place a moistened paper tissue over the comb, and cover the upper end of the assembly with cling film to prevent the gel from drying out.

3.3. Running the Gel

1. Remove the bulldog clips and adhesive tape from the long edges of the gel assembly. Specifically designed clamps may be used to remain in place during electrophoresis. Also remove all components used to seal the bottom edge, e.g., spacer, clips, tape, casting tray, and so forth.

2. Remove the comb and secure the plate assembly into the sequencing apparatus using either bulldog clips, or the support bar and screws provided.

3. Make up the recommended quantity of 1X TBE buffer (about 1100 mL for a 21 × 50 cm gel), and pour into the upper buffer chamber to about 1 cm from the top. Using a 50-mL syringe and needle, squirt some TBE into the sample wells of the gel to rinse away any unpolymerized acrylamide. If a shark's tooth comb is employed, the straight edge of the gel should be similarly rinsed. The comb should then be washed to remove any acrylamide or urea, and reinserted so that the points of the teeth just pierce the gel by about 1 mm. Once this is done, the comb should not be moved subsequently, since leakage of samples between wells may result. The comb can be secured in place with two miniature bulldog clips if desired.

4. Check that no buffer is leaking from the upper chamber into the lower one (plug any gaps with molten agarose if necessary). Pour the remaining buffer into the lower buffer tank ensuring the electrodes are immersed. If a bottom spacer was used during pouring, there may be an air space at the bottom of the gel. This can be removed by squirting TBE into the space with a syringe and attached needle, which has been bent at 45° halfway along its length.

5. Load 5 μL of the formamide indicator dye, which is used to stop the sequencing reactions, into a few of the wells, and run the gel at an appropriate voltage (e.g., 2000 V for an 21 × 50 cm 8% gel; ~50 W) for 20–60 min until the temperature

stabilizes at 55°C. Temperature is best measured by a temperature indicator attached to the outer glass plate. It is important not to let the temperature exceed 65°C, since this may hydrolyze the gel or cause the glass plates to crack. During this time, the level of the buffer in the upper chamber may drop owing to expansion of the apparatus on warming and should be topped up as necessary.

6. Denature the sequencing reaction samples into single strands by heating to 95°C for 3 min. If a microtiter plate has been used for the reactions, incubation should be at 80°C for 10 min to avoid the risk of melting the plate. Immediately plunge them into ice to prevent reannealing.

7. Turn off the power supply, and rinse the loading wells once more with TBE. Load 5 μL of each of the four termination reactions from each template into adjacent wells of the gel. If a shark's tooth comb has been used, samples may be applied with a Pipetman. Otherwise, a 5- or 10-μL syringe with a 28- or 30-gage needle may be necessary. If a syringe is used, the needle should be rinsed well between each sample. If sample migration across sample wells is observed, "staggered" loading may be employed in which the gel is run for 2–3 min between each loading of a complete set of four reactions. If there are spare slots in the gel, it is also advisable to load a mol-wt marker at one edge (or two different ones at each edge), both to assist in identifying the position of the sequence, and to orient the final autoradiograph.

8. Reconnect the power supply and run the gel at 50°C until the sample dyes have migrated the required distance. As a guide, bromophenol blue migrates with a DNA fragment of approx 26 nucleotides, and xylene cyanol with a fragment of approx 106 nucleotides, in a 6% gel.

3.4. Gel Drying and Autoradiography

1. At the end of the run, disconnect the power supply and remove the gel. Rinse buffer off the gel, and discard buffer from the gel apparatus into a sink designated for liquid radioactive waste. Remove any clamps or clips that secured the plates together during the run.

2. Remove the siliconized plate by gently prising the plates apart at one end. The gel should adhere to the other plate, but care needs to be taken, since occasionally the gel may adhere to the siliconized plate or partly to both plates (*see* Note 7).

3. Cut a piece of Whatmann 3MM paper to the appropriate size, and gently lay it on top of the gel. Rub the back of the Whatmann paper with a paper towel, and then peel it away from the glass plate. The gel will remain stuck to the Whatmann paper.

4. Cover the gel with cling film, smoothing out creases and air bubbles with a paper towel.

5. Dry the gel in a slab gel dryer at 80°C for 30–120 min until the gel is dry.

6. Remove the cling film (this is essential with ^{35}S because it is such a weak β emitter, but not necessary with ^{32}P), and autoradiograph the gel against a high-speed X-ray film in a suitable cassette. If ^{32}P is used, an intensifying screen should be used and the cassette incubated at –70°C. With ^{35}S, incubation can be at room temperature, without a screen. Exposure times vary from about 1 to 10 d

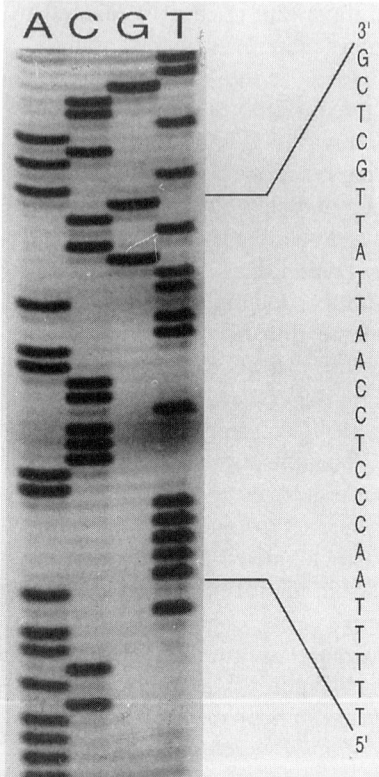

Fig. 2. Autoradiograph of a sequencing gel. The sequence is derived from a single-stranded template isolated from a PCR product amplified from the human factor VIII gene.

depending on the type and age of the radioactivity, and the quantity of starting DNA template.

3.5. Analysis of Gels

Figure 2 shows an autoradiograph of a sequencing gel. Each vertical position in the gel is occupied by a band in one of the four lanes representing each base. The sequence is read from the bottom of the gel (smaller fragments, closer to primer) to the top (longer fragments, further from primer).

Sometimes the bands on the autoradiograph form a curved pattern across the gel instead of lying in a straight line. This is known as gel "smiling," and occurs when samples near the center of the gel run faster than those at the edges. It is because of the more rapid dissipation of heat near the edges. Features incorporated into the gel apparatus may substantially reduce this problem by the use of a thermostatic

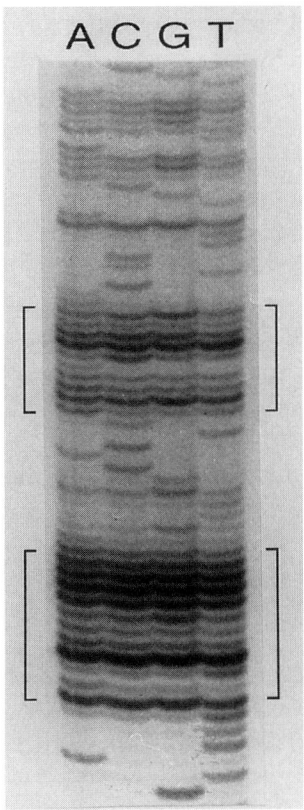

Fig. 3. Areas of compression (brackets) at regions of secondary structure in the DNA template.

plate adjacent to the gel plate, or the design of a buffer chamber that extends over the entire area of the gel enabling the dissipation of heat by convection.

Adjacent bands of DNA may become compressed and appear across all four lanes of the sequencing gel (Fig. 3). This is owing to intrastrand secondary structure in the DNA arising at regions of dyad symmetry, especially those with a high G + C content. They occur despite protective steps to minimize their formation (*see* Section 1). This artifact is commonly observed with double-stranded DNA templates (e.g., plasmids and PCR products) and is less of a problem with single-stranded templates. DNA sequencing analogs, such as 2'-deoxyinosine-5'-triphosphate (dITP) and 7-deaza-2'-deoxyguanosine-5'-triphosphate (7-deaza-dGTP), which pair more weakly than the conventional bases, may help to resolve compressions *(1,2)*. Alternative methods are to use a single-stranded DNA-binding protein (Amersham International, Amersham, UK 70032) or to sequence both strands of the DNA.

The number of bases that can be resolved from a single load can be increased from 200–250 to around 600 by running a buffer gradient gel *(1)*. This involves preparing two gel mixes with the same acrylamide and urea concentrations, one of 0.5X TBE and the other with 5X TBE. About a third of the total gel volume is initially prepared by taking up equal volumes of the 0.5X TBE mix followed by the 5X TBE mix into the same pipet, and then introducing a few air bubbles to mix the solutions at the interphase. This solution is poured into the mold, which is then filled with only 0.5X solution. The success of the gradient formation can be monitored by the addition of 0.05 mg/mL of bromophenol blue to the 5X solution. The increasing ionic concentration in the bottom third results in compression of the lower-mol-wt fragments at the bottom of the gel and allows better resolution of high-mol-wt fragments near the top. An alternative technique to achieve the same result is to use a wedge-shaped gel, which is poured using spacers that are 0.6–0.75 mm thick at the bottom and taper to 0.25 mm at the top *(3)*. However, because these gels take longer to dry and are more prone to cracking, buffer gradient gels are the preferred choice in most laboratories.

4. Notes

1. Unpolymerized acrylamide is a neurotoxin. Gloves and mask should be worn, and care should be taken when handling acrylamide powder or solutions. Depending on the quality of the acrylamide, it may be advisable to stir the acrylamide with monobed resin (MB-1) to remove contaminating metal ions prior to filtration.
2. Leaks and air bubbles constitute the most problematic aspect of successful gel pouring. The key to avoiding these is thorough cleaning of gel plates and spacers, and is best performed immediately after use. Clean the plates by scrubbing with a nonabrasive detergent and rinsing with tap water. Wipe the plates with a dry paper towel, followed by a second towel, which has been moistened with ethanol. After the initial cleaning with detergent and drying, the siliconized plate should additionally be wiped with a paper towel wetted with siliconizing solution, allowed to dry, and wiped with deionized water. The gel kit and any reusable items used for securing the plates for gel pouring should be washed with a mild detergent and warm water.
3. It is advisable to distinguish the two sides of each glass plate to ensure that the same side is always used for the inner ("gel") and outer sides, and that it is always the same side of one of the plates that is siliconized. The design of some manufacturer's apparatus may ensures this. In other cases, temperature decals, sticky tape, or a permanent marker may be used to identify the outer side.
4. Some gel kits provide alternative sealing methods, e.g., specifically designed clamps for the sides and a casting tray and protocol that uses an acrylamide-saturated paper strip to seal the bottom edge *(4)*. These are usually easier and more reliable than clips and tape.

5. The concentration of acrylamide to be used depends on the distance of the sequence to be resolved from the sequencing primer. A 6% gel is suitable for reading between 25 and 400 nucleotides from the primer. Higher concentrations (12–20%) may be used for sequences within 50 nucleotides, and lower concentrations (4 or 5%) for >400 nucleotides.

6. Some protocols recommend degassing the urea/TBE/acrylamide gel mix immediately prior to use to reduce the chance of air bubbles forming when pouring the gel, but this step is not essential.

7. Many protocols recommend fixing the gel in 10% acetic acid and 10% methanol for 15 min on one glass plate before transfer to the Whatmann paper. This procedure removes urea and decreases the time required for drying. However, it also increases the chances of the fragile gel tearing or folding back on itself, and so is probably best omitted.

References

1. Biggin, M. D., Gibson, T. J., and Hong, G. F. (1983) Buffer gradient gels and ^{35}S label as an aid to rapid DNA sequence determination. *Proc. Natl. Acad. Sci. USA* **80,** 3963–3965.

2. Mizusawa, S., Nishimura, S., and Seela, F. (1986) Improvement of the di-deoxy chain termination method of DNA sequencing by use of deoxy-7-deazaguanosine triphosphate in place of dGTP. *Nucleic Acids Res.* **14,** 1319.

3. Reed, A. P., Kost, T. A., and Miller, T. J. (1986) Simple improvements in ^{35}S dideoxy sequencing. *BioTechniques* **4,** 306.

4. Wahls, W. P. and Kingzette, M. (1988) No runs, no drips, no errors: a new technique for sealing polyacrylamide gel electrophoresis apparatus. *BioTechniques* **6,** 308.

2

Purification of PCR Products
from Agarose Gels for Direct Sequencing

Frank C. Brosius III, Lawrence B. Holzman, and Xinan Cao

1. Introduction

The advent of direct sequencing of polymerase chain reaction (PCR) products has permitted extremely rapid analysis of DNA mutants and cDNA clones. However, direct PCR sequencing has been problematic for a number of technical reasons, including the presence of impurities and excess oligonucleotide primers used for the PCR amplifications *(1–4)*. Therefore, a number of protocols have been devised that address these technical issues, and allow efficient sequencing of either conventional double-stranded PCR products or asymmetrically amplified single-stranded products (e.g., *1–4*). Many of these protocols are described in detail in this volume.

An unrelated, but equally frustrating obstacle to the sequencing of PCR products arises when multiple PCR products are obtained in a single reaction. This can occur because of nonspecific amplification, when using degenerate primers or low stringency annealing conditions, or when the target cDNA is of low abundance. Multiple PCR products can also arise when cDNAs representing several alternatively spliced mRNAs of the sequence of interest are present. In each of these instances, separation and purification of the desired PCR products is necessary before sequencing is possible. The purpose of this chapter is to describe a simple, low-cost method by which multiple PCR products can be purified from agarose gels for direct sequencing *(5)*. We have utilized a modification of the procedure of Heery et al. *(6)*, which describes a simple, low-speed centrifugation of agarose slices to obtain the PCR products. After phenol-chloroform extraction, the products are ready for DNA sequencing using either of the PCR oligonucleotides as a sequencing primer. A few PCR products cannot be sequenced by this method, but can be sequenced after a further purification step, which is also detailed below.

From: *Methods in Molecular Biology, Vol. 65: PCR Sequencing Protocols*
Edited by: R. Rapley Humana Press Inc., Totowa, NJ

2. Materials

All chemicals should be of molecular biology grade. All solutions should be made with double-distilled or deionized water. Reagents for sequencing are available as kits.

2.1. Agarose Gel Electrophoresis

1. Ultrapure agarose (Gibco BRL, Life Technologies, Gaithersburg, MD): Use of a low-percentage agarose gel (<1.0%) appears to yield a better-quality sequence (*see* Note 1).
2. 1X TAE buffer: 0.04M Tris-acetate, 0.001M EDTA: This can be made up in bulk as 20X or 50X stock and stored at room temperature (*see* Note 2).
3. Ethidium bromide (EtBr) solution: EtBr is a mutagen. Therefore, adequate safety precautions should be used when handling this reagent (*see* Note 3).
4. 10X loading buffer: 0.25% bromophenol blue, 0.25% xylene cyanol FF, 30% glycerol in ddH$_2$O.
5. UV transilluminator.

2.2. Isolation of PCR Products from Individual Gel Slices

1. Silanized glass wool: Glass wool can be silanized by submerging glass wool in a 1:100 dilution of a siliconizing agent, such as Sigmacoat (Sigma, St. Louis, MO) or Prosil 28 (VWR, Chicago, IL) for 15 s, and then rinsing extensively with distilled water, followed by autoclaving for 10 min. Silanized glass wool may be stored at room temperature indefinitely.
2. 1.5-mL Eppendorf centrifuge tubes: Tubes should be capless or have caps and hinges removed.
3. Phenol/chloroform/isoamyl alcohol 25:24:1 solution: Make by mixing 25 vol of buffered phenol with 24 vol of chloroform and 1 vol of isoamyl alcohol. Store under 50 mM Tris-HCl, pH 8.0, at 4°C. Phenol can cause severe burns to skin and mucosal membranes (*see* Note 3).
4. 5M Ammonium acetate.
5. 100% Ethanol.
6. TE: 10 mM Tris-HCl, 1 mM EDTA, pH 8.0.

2.3. PCR Product Denaturation and DNA Sequencing

1. 1M NaOH.
2. 3M Sodium acetate, pH 5.2.
3. 5M Ammonium acetate.
4. 100% Ethanol.
5. 5X Sequenase buffer: 200 mM Tris-HCl, pH 7.5, 100 mM MgCl$_2$, 250 mM NaCl (available in the Sequenase kits from United States Biochemical, Cleveland, OH).
6. 10 pmol 5'- or 3'-oligonucleotide primer used for PCR amplification (*see* Note 4).
7. 5X labeling mix (dGTP): 7.5 μM dGTP, 7.5 μM dCTP, 7.5 μM dTTP. There is no dATP in this mix. We often obtained better sequence by using 7.5 μM deaza-dGTP in place of dGTP (*see* Note 5).

8. Sequenase version 2.0 modified T7 DNA polymerase (14 U/µL): Dilute eight-fold in cold enzyme dilution buffer just before use (*see* Note 6).
9. Enzyme dilution buffer: 10 m*M* Tris-HCl, pH 7.5, 5 m*M* DTT, 0.5 mg/mL BSA. Also available in Sequenase kits.
10. 0.1*M* DTT: Store at –20°C.
11. Dideoxy (dd) G termination mix: 80 µ*M* dGTP, 80 µ*M* dATP, 80 µ*M* dCTP, 80 µ*M* dTTP, 8 µ*M* ddGTP, 50 µ*M* NaCl. We often obtained better sequence by using 7.5 µ*M* deaza-dGTP in place of dGTP in each of the termination mixes (*see* Note 5). The termination mixes (either with dGTP or deaza-dGTP) are available as part of the Sequenase kits (United States Biochemical).
12. Dideoxy (dd) A termination mix: 80 µ*M* dGTP, 80 µ*M* dATP, 30 µ*M* dCTP, 80 µ*M* dTTP, 8 µ*M* ddATP, 50 m*M* NaCl.
13. Dideoxy (dd) T termination mix: 80 µ*M* dGTP, 80 µ*M* dATP, 80 µ*M* dCTP, 80 µ*M* dTTP, 8 µ*M* ddTTP, 50 m*M* NaCl.
14. Dideoxy (dd) C termination mix: 80 µ*M* dGTP, 80 µ*M* dATP, 80 µ*M* dCTP, 80 µ*M* dTTP, 8 µ*M* ddCTP, 50 m*M* NaCl.
15. Formamide stop solution/gel loading buffer: 95% formamide, 20 m*M* EDTA, 0.05% bromophenol blue, 0.05% xylene cyanol FF. Also available in Sequenase kits.
16. [$\alpha^{35}S$]-dATP, 1000 Ci/mmol (Amersham, Arlington Heights, IL)
17. Sequencing gel: 6% denaturing acrylamide gel, 0.4 mm thick.

2.4. Alternate PCR Product Purification Method

1. Acid-phenol reagent: 4*M* guanidinium thiocyanate, 25 m*M* sodium citrate, pH 7.0, 0.5% sarcosyl, 0.2*M* sodium acetate, pH 4.0. This is the reagent used by Chomczynski and Sacchi for RNA harvest *(7)* (*see* Note 7).
2. 100% Ethanol.

3. Methods
3.1. Agarose Gel Electrophoresis

1. PCR reactions can be run conventionally. We have often run 100-µL reactions in order to generate approx 0.5 µg of each PCR product (band) for sequencing (*see* Note 8).
2. Pour a conventional 0.8% agarose, 1X TAE gel containing 0.5 µg/mL ethidium bromide. Use a comb that will generate wells that can accommodate 111-µL vol (*see* Note 1).
3. Add 1/10 vol of agarose gel loading buffer to PCR reaction, and load into wells.
4. Electrophorese at 100 V until adequate separation of PCR bands is achieved.
5. Puncture 1.5-mL Eppendorf centrifuge tubes with a 25-gage needle at the bottom. Plug bottom of tubes with loosely packed silanized glass wool (Fig. 1).
6. Place gel directly onto UV transilluminator. Excise bands of interest with a fresh scalpel blade or razor blade, and place into punctured and plugged 1.5-mL Eppendorf tubes (*see* Fig. 1). Excise the PCR band in the smallest gel slice possible.

3.2. Isolation of PCR Products from Individual Gel Slices

1. Place 1.5-mL tube with gel slice directly into a second empty 1.5-mL capless tube, and place the entire assemblage inside a 15-mL plastic centrifuge tube.

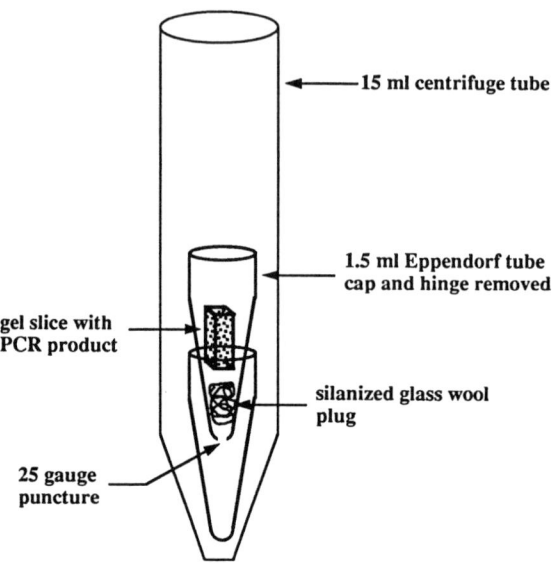

15 ml centrifuge tube

1.5 ml Eppendorf tube
cap and hinge removed

gel slice with
PCR product

silanized glass wool
plug

25 gauge
puncture

Fig. 1. Schema for isolation of PCR products by low-speed centrifugation of a conventional agarose gel slice.

2. Centrifuge at room temperature for 10 min at 400*g* in a tabletop centrifuge (swinging bucket); *g*-force should be calculated using a radius measured to the silanized glass plug (*see* Note 9).
3. Extract the centrifugate with an equal volume of phenol/chloroform. Add 2/5 vol of 5*M* ammonium acetate to the centrifugate. Ethanol-precipitate by adding 2.5 vol of cold 100% ethanol and placing in –80°C freezer for 15–30 min (*see* Note 10).
4. Centrifuge at 13,000*g* in a microfuge for 30 min.
5. Redissolve pellet in 20-µL of TE. If the PCR product needs to be reamplified, 1 µL of this stock solution can be used for subsequent PCR reactions (*see* Note 11).
6. Centrifuge this 20-µL stock solution in a microfuge for 3–5 min to pellet any remaining agarose. Carefully transfer supernatant to another tube for DNA sequencing (*see* Note 12).

3.3. PCR Product Denaturation and DNA Sequencing

1. To 14 µL of PCR product stock solution, add 2.8 µL of 1*M* NaOH at room temperature for 5 min. Vortex.
2. Place tube in ice H_2O bath, and neutralize with 7 µL 5*M* ammonium acetate, pH 5.2. Vortex.
3. Add 60 µL 100% ethanol. Invert several times and place in –80°C freezer for 5 min.
4. Centrifuge in a microfuge at 13,000*g* for 10 min at room temperature. Wash pellet once with 100 µL of cold 70% ethanol. Aspirate and dry pellet. Resuspend in 7 µL ddH_2O.

5. Add 2 µL 5X Sequenase buffer and 1 µL (10 pmol) of either the 3'- or 5'-primer used for PCR amplification (*see* Note 13).

6. Anneal primer to PCR product at 54°C for 2 min, and then allow to cool to room temperature slowly (*see* Note 14).

7. While PCR product–primer mix is cooling, dilute Sequenase enzyme 1:8 in ice-cold enzyme dilution buffer (*see* Note 6 *[8]*).

8. Also dilute 5X labeling mix fivefold with ddH$_2$O.

9. Pipet 2.5 µL of the appropriate termination mix into tubes labeled G, A, T, and C (ddGTP termination mix into tube "G," and so forth).

10. After PCR product–primer mix is cooled, add:
    a. 0.1*M* DTT   1.0 µL
    b. Diluted labeling mix   2.0 µL
    c. [$\alpha^{35}S$] dATP   0.5 µL
    d. Diluted Sequenase   2.0 µL
    Mix thoroughly and incubate for 5–10 min at room temperature (*see* Note 15).

11. Heat tubes with termination mixes to 37°C.

12. Add 3.5 µL of the labeling reaction mixture to the tubes with the termination mixes, mix well, and return to 37°C for 20–30 min (*see* Note 15).

13. Add 4 µL of formamide stop solution/gel loading buffer to each tube, mix well, and store at 4°C if sequencing gel is to be run the same day or at –20°C if gel is to be run later.

14. Pour a 6% sequencing gel (*see* Chapter 1).

15. Heat mixes to 95°C for at least 5 min before loading on gel.

We have successfully employed this method with multiple unrelated primers and PCR products, and generally obtained at least 200 bp of readable sequence. An autoradiograph of a typical sequencing gel using this method is shown in Fig. 2. Some of the common problems we have encountered with this method are detailed in Note 16.

With a few products, the extent of readable sequence is sometimes limited to approx 150 bp, and occasionally, a primer/PCR product pair produces an unreadable sequence. In the latter instance, a simple purification of the DNA product has led to generation of readable DNA sequence of at least 100 bp and is presented in the alternate purification protocol below.

3.4. Alternate PCR Product Purification Method

1. Dissolve the PCR product (again at least 0.5 µg) in 100 µL of denaturing solution *(7)*.

2. Add 100 µL of phenol. Vortex and centrifuge in a microfuge at 13,000*g* for 5 min. Under these conditions, the DNA remains in the interphase and the organic phase (*see* Note 7).

3. Carefully remove and discard the aqueous phase.

4. Add 200 µL of TE. Vortex and centrifuge in a microfuge at 13,000*g* for 5 min. Remove and retain the aqueous phase. Much of the DNA will now be retained in the aqueous phase (*see* Note 7).

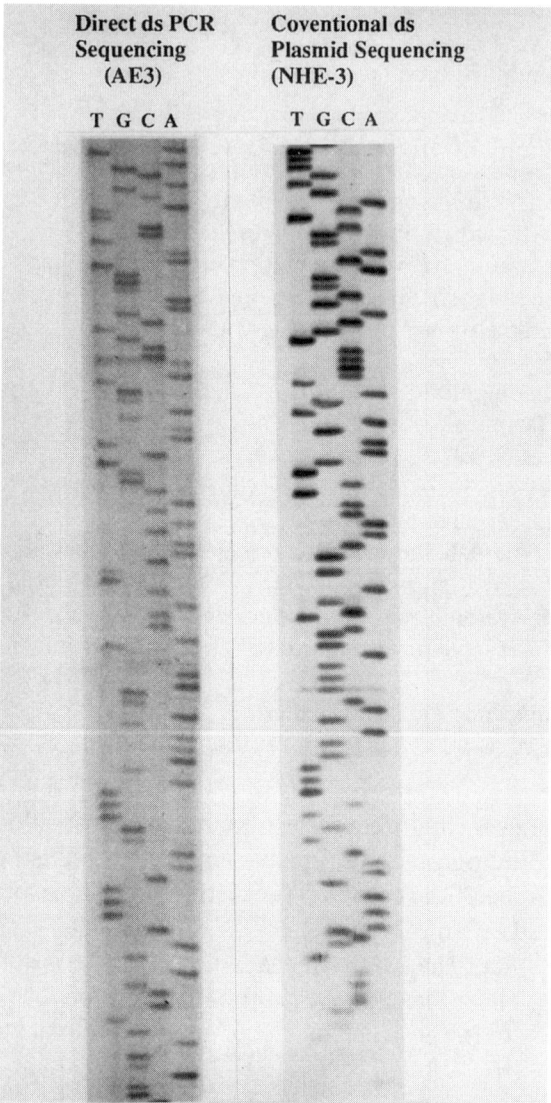

Fig. 2. Typical result of PCR product sequencing using the described protocol vs conventional DNA sequence obtained from a double-stranded plasmid template. The direct PCR sequence gel was exposed to film for 40 h, whereas the conventional plasmid sequence gel was exposed overnight. "AE3" signifies anion exchanger 3, and "NHE-3" signifies sodium proton exchanger 3.

5. Add 22 μL of 3M sodium acetate to the aqueous phase and 450 μL of 100% ethanol. Let sit at room temperature for 10–15 min. Vortex and centrifuge in a microfuge at 13,000g for 15 min.
6. Resuspend pellet in 14 μL TE, and return to Section 3.3., step 1.

We do not know which factors are most important for the success of this alternate protocol. It is assumed that further purification of PCR products enhances sequence reaction efficiency. It is also possible that improved denaturation of the PCR products during extraction into the organic phase permits better primer annealing.

4. Notes

1. The use of low-percentage agarose gels (0.8% is our usual concentration) is associated with better PCR sequence results. This may result from decreased agarose contamination of the purified PCR product, but this has not been formally tested. Since most of the PCR products used for sequencing are between 500 and 1000 bp in length, this agarose concentration does not yield ideal separation of the various bands. When separation of PCR products is not optimal or when PCR product yields are low (<0.5 μg), we use 2% Metaphor agarose (FMC Bioproducts, Rockland, ME) to obtain better resolution and increase PCR yield by reamplification. These reamplified products are then purified again on 0.8% conventional agarose gels. We have not yet attempted to sequence PCR products directly after isolation from a Metaphor agarose gel.
2. TAE can either be made up as a 20X or 50X stock, or 1X TAE can be made up in bulk in a carboy and used directly. We have made up 15-L solutions of 1X TAE, which are stable for several months at room temperature.
3. EtBr and phenol-containing solutions should be made up in a fume hood or other confined space. Personnel preparing EtBr solutions should wear adequate protective clothing. Stock solutions of 10 mg/mL EtBr should be stored in a light-protected glass container at 4°C. Such solutions are stable for several years.
4. One of the advantages of this protocol is that the primers used for PCR amplification can also be used for DNA sequencing. We conventionally use 20 mers, which have a G + C content of 50% (T_m ~ 60°C). Use of primers with different T_ms should work equally well, but may require appropriate changes in the annealing temperature of primer and the DNA for sequencing (*see* Section 3.3., step 6).
5. Deaza-dGTP helps prevent GC compression on sequencing gels and can be used in place of dGTP in all the sequencing reagents. The deaza-dGTP-containing reagents are also available in kit form from US Biochemicals (Cleveland, OH).
6. For routine double-stranded DNA sequencing, we have utilized glycerol enzyme dilution buffer, which can be stored for extended periods at –20°C. The use of this reagent, however, requires the use of glycerol-tolerant DNA sequencing gels, which is described in the Sequenase protocol book. Although we have not used glycerol enzyme dilution buffer and glycerol-tolerant gels for direct sequencing of PCR products, such modifications should not pose difficulties and would help save expensive sequencing enzyme.
7. This acid-phenol reagent is commonly used for RNA harvest, but because DNA is partitioned into the organic phase, it can be utilized to separate and purify plasmid and genomic DNA as well as RNA *(9)*. This extra purification step apparently provides PCR product purification that is superior to that which can be obtained with phenol-chloroform extraction alone and, therefore, improves the sequence from difficult templates. Our method uses a similar method, except for the back-extraction and final precipitation step. Chomczynski's protocol recommends direct precipita-

tion from the organic phase and interphase using ethanol *(9)*. We have not yet tested this direct precipitation step for PCR products that are used for direct sequencing.

8. We routinely estimate PCR product yield by the relative intensity of the PCR band compared to mol-wt ladder bands when UV transilluminated. It is not necessary to measure precisely the PCR product yield. As an alternative to reamplification, multiple separate reactions could be run, and the final PCR products pooled.

9. For example, a *g*-force of 400 is obtained at 1500 rpm in a Beckman GS-6R tabletop centrifuge using a GH-3.8 rotor. It will be necessary to check the specifications of each low-speed centrifuge and rotor to determine the correct rpm.

10. We use ammonium acetate at this step, because some of our less abundant bands need to be reamplified. DNA precipitated with ammonium acetate provides better PCR amplification than does DNA precipitated with sodium acetate *(10)*.

11. Since we use this protocol to help distinguish multiple PCR bands, the lower abundance bands often yield <0.5 μg of DNA. We have found that such small amounts of DNA do not provide adequate template for Sequenase-based sequencing. Therefore, we simply reamplify these products with another 30 cycles of PCR and isolate the product for sequencing again as described in the protocol. We have not tested whether a PCR-based cycle sequencing method would work well with these rare products and, therefore, obviate the need for reamplification.

12. The purpose of this step is to remove any residual agarose or other insoluble contaminants that may inhibit sequencing. It is probably not essential if the extraction and precipitation steps are carefully performed.

13. Either of the primers used to amplify the PCR product of interest can be used for sequencing. The amount of primer used in this step (10 pmol) is 5× that specified in the published Sequenase protocol *(8)*.

14. Many protocols for sequencing of PCR products call for rapid cooling of template and primer after heating to the annealing temperature in order to prevent reannealing of the two PCR strands (e.g., *2,3*). For reasons that are not clear, the slow cooling suggested in this protocol yielded more intense sequence than did more rapid cooling to room temperature. We have not attempted snap-cooling to 4°C or −70°C as recommended by some protocols, and this may increase intensity of the sequence ladder.

15. The labeling reaction duration of 5–10 min is somewhat longer than that recommended in the Sequenase protocol book (2–5 min) *(8)*. Similarly, the termination reaction time of 20–30 min is extended.

16. The major problems encountered with this protocol are high background, relatively low intensity of DNA sequence radioactivity, and frequent "hard stops" as evidenced by bands in each lane at the same mobility. Often the first two problems are overcome by repurification using the alternate purification protocol described in Section 3.4. Also, a 2–3 d exposure to film is sometimes necessary to obtain clearly readable sequence. Also, some of the problems with "signal-to-noise" can be diminished with the use of larger amounts (>1 μg) of starting PCR product. Occasionally, use of a certain primer fails to produce readable sequence despite having a calculated T_m and base composition identical to other primers that work well. In that case, use of an internal primer for sequencing may be necessary. The generation of hard stops remains something of a mystery. As shown in Fig. 3, we have obtained intense

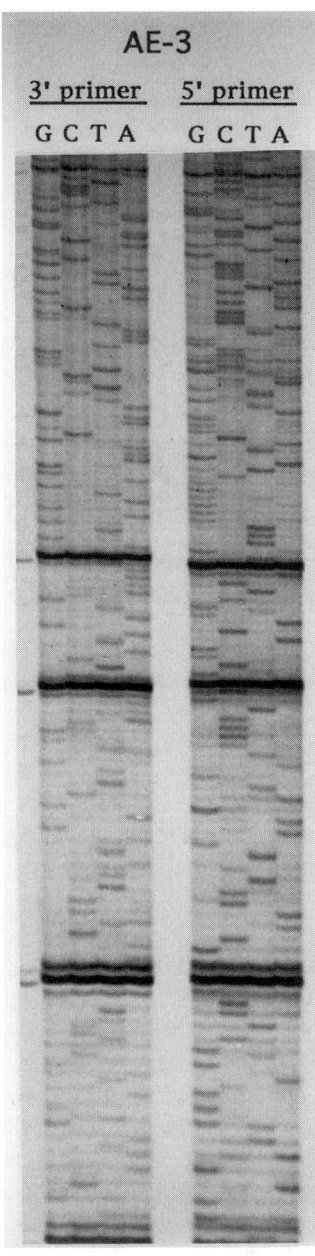

Fig. 3. Direct PCR sequence gels showing intense bands in all four lanes indicative of "hard stops" owing to premature termination of DNA synthesis. Curiously, bands of the same mobility were seen whether the 5'- or 3'-PCR primer was used as a sequencing primer, suggesting that some factor other than secondary structure or reannealing of the PCR product was responsible. With this particular PCR product preparation, relatively poor sequence ladders were generated, suggesting poor template purification.

bands in each sequencing lane while sequencing the same cDNA with either 3'-
or 5'-primers. Since the sequence reaction with the 3'-primer proceeds from the
opposite end of the PCR product than that with the 5'-primer, it seems unlikely
that secondary structure or reannealing of the PCR template is the cause of this
phenomenon. Perhaps a contaminating PCR product or other contaminant allows
for nonspecific priming and extension up to a region in which the Sequenase
prematurely terminates. In some cases, these hard stops may be eliminated by
enhanced purification.

Several other protocols for direct sequencing of PCR products in low melt
agarose gel slices have been recently published *(11,12)*. These protocols have not
been utilized by the authors, but should be considered simple alternatives to the
protocol described herein.

Acknowledgment

This work was supported, in part, by a Veterans Administration Merit
Review award and a National Institutes of Health program project award
(HL18575 Project 2) to F. C. B.

References

1. Dorit, R. L., Ohara, O., and Hwang, C. B.-C. (1991) Direct DNA sequencing of
   PCR products, in *Current Protocols in Molecular Biology,* vol. II (Ausubel, F.
   M., Brent, R., Kingston, R. E., Moore, D. D., Seidman, J. G., Smith, J. A., and
   Struhl, K. S., eds.), John Wiley, New York, pp. 15.2.1–15.2.11.
2. Phear, G. A. and Harwood, J. (1994) Direct sequencing of PCR products. *Methods
   Mol. Biol.* **31,** 247–256.
3. Cassanova, J.-L., Pannetier, C., Jaulin, C., and Kourilsky, P. (1990) Optimal con-
   ditions for directly sequencing double-stranded PCR products with Sequenase.
   Nucleic Acids Res. **18,** 4028.
4. Rao, V. B. (1994) Direct sequencing of polymerase chain reaction-amplified
   DNA. *Anal. Biochem.* **216,** 1–14.
5. Cao, X., and Brosius, F. C., III. (1993) Direct sequencing of double-stranded
   PCR products isolated from conventional agarose gels. *BioTechniques* **15,**
   384–386.
6. Heery, D. M., Gannon, F., and Powell, R. (1990) A simple method for subcloning
   DNA fragments from gel slices. *Trends Genet.* **6,** 173.
7. Chomczynski, P. and Sacchi, N. (1987) Single-step method of RNA isolation by
   acid guanidinium thiocyanate-phenol-chloroform extraction. *Anal. Biochem.* **162,**
   156–159.
8. United States Biochemical Corporation (1992) *Step-by-Step Protocols for DNA
   Sequencing with Sequenase version 2.0 T7 DNA Polymerase,* 6th ed., Cleveland, OH.
9. Chomczynski, P. (1993) A reagent for the single-step simultaneous isolation
   of RNA, DNA and proteins from cell and tissue samples. *BioTechniques* **15,**
   532–534.

10. Coen, D. M. (1992) Quantitation of rare DNAs by PCR., in *Current Protocols in Molecular Biology*, vol. II (Ausubel, F. M., Brent, R., Kingston, R. E., Moore, D. D., Seidman, J. G., Smith, J. A., and Struhl, K. S., eds.), John Wiley, New York, pp. 15.3.1–15.3.6.
11. Khorana, S., Gagel, R. F., and Cote, G. J. (1994) Direct sequencing of PCR products in agarose gel slices. *Nucleic Acids Res.* **22(16),** 3425–3426.
12. Trewick, S. A. and Dearden, P. (1994) A rapid protocol for DNA extraction and primer annealing for PCR sequencing. *BioTechniques* 17, 842–844.

3

Enzymatic Fluorescence and Biotin Labeling of Primers for PCR Sequencing

Gabor L. Igloi

1. Introduction

The emergence of cycle sequencing *(1)* as a powerful alternative to conventional isothermal methods has facilitated the manipulations involved in sequencing protocols in general, and is of value for the sequence analysis of PCR products, in particular. Using radioactive labeling techniques, it has generally been accepted that 5'-terminal phosphorylation of the primer prior to the sequencing reaction *(2)* is the most reliable approach to obtaining sequence information. However, the combination of nonisotopic detection with cycle sequencing offers several alternatives. The incorporation of the fluorescent label into the nascent DNA has been approached from several angles. At present, one may consider three sources of fluorescence (Fig. 1), irrespective of the principle governing the design of the sequencing hardware:

1. Label attached at the 5'-terminus of the primer *(3)*.
2. Label incorporated as an appropriately chemically modified dideoxy-terminating nucleotide *(4)*.
3. Internal labeling of the growing chain using F-dATP during the elongation reaction *(5)*.

In practice, the source of the label is governed to a large extent by the nature of the available automated sequencer or, more specifically, by the principle involved in the fluorescence detection. To date, two types of data acquisition are of commercial significance: (1) a scanning, multiple-wavelength approach, permitting a "four dye/one lane" analysis and (2) a single-wavelength "one dye/four lane" system. The advantages and disadvantages of these alternatives will not be discussed at this point. However, it is evident (and confirmed in

From: *Methods in Molecular Biology, Vol. 65: PCR Sequencing Protocols*
Edited by: R. Rapley Humana Press Inc., Totowa, NJ

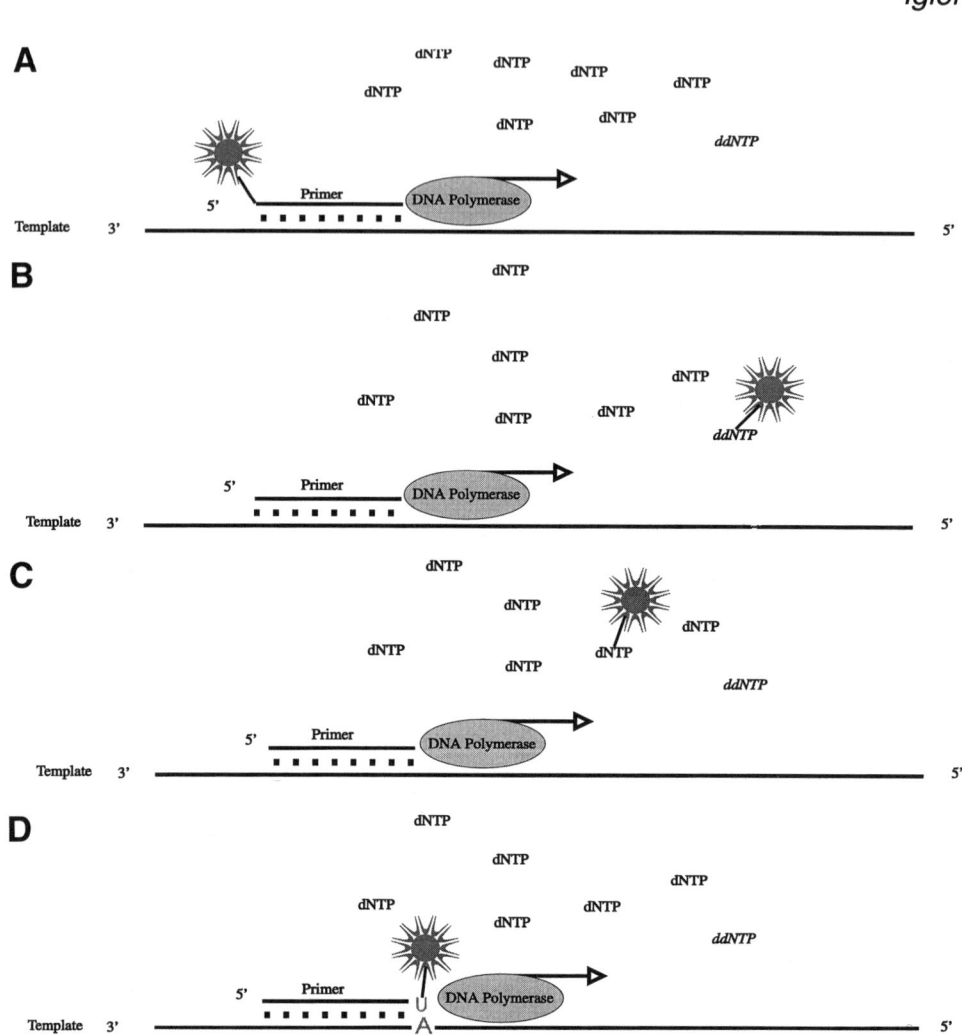

Fig. 1. Source of label during fluorescent DNA sequencing. The fluorescent tag is indicated schematically by a star.

practice) that it is not practicable to synthesize four differently labeled primers for the "four dye/one lane" system when large-scale sequencing projects, involving large numbers of walking primers, are being carried out. In this case, the use of labeled terminator nucleotides is the method of choice despite their expense, the additional steps in removing the unincorporated label, and the limited amount of sequence information that is routinely obtainable. On the other hand, for "one dye/four lane" sequencers, the labeling of primers can be straightforward, particularly if fluorescein is to be the chromophore. In this case, the availability of the corresponding phosphoramidite provides a convenient, if expensive, route to a 5'-terminal label during primer synthesis. Internal labeling with fluorescent dATP *(5)* is a useful alternative, but the cycle sequencing protocol is more effective with fluorescein-dCTP *(6)*. However, the fact that during the polymerization process the 3'-terminally incorporated fluorescein-dNTP does not block elongation (by steric or other effects) is of significance for the method described here.

The large arsenal of DNA-modifying enzymes available to the molecular biologist has been used for decades for the radioactive labeling of nucleic acids. Using the activity of one of these, terminal deoxynucleotidyltransferase, it is possible to tag postsynthetically a primer with either a fluorescent label or with biotin. The procedure described in this protocol gives rise to an oligonucleotide elongated 3'-terminally with a ribonucleotide bearing the appropriate label and whose free 3'-OH is accepted as a substrate for the template-dependent DNA polymerase (Fig. 1D). The major product of this enzymatically well-characterized manipulation *(7,8)* is an oligodeoxynucleotide chain that has been elongated by a single labeled ribonucleotide. The terminal addition must, for the applications envisaged here, be complementary to the target sequence. Since, at present, commercially available labeled ribonucleotides are limited to the UTP analogs (carrying the label at the 5-position of the base), the design of primers must take this limitation into account (*see* note added in proof). Thus, for convenience, primers may be selected by a variety of computer programs to terminate in a 3'-T. The primer is synthesized chemically lacking this residue and, for fluorescence-based sequencing, a fluorescein-rUMP is then added enzymatically. Analogously, the enzymatic introduction of a 3'-terminal biotin-rUMP molecule into a PCR primer permits the subsequent isolation of the biotinylated DNA single strand for, e.g., direct solid-phase sequencing of PCR products (*see* Chapter 9). As far as we are aware, this approach offers the unique possibility of enzymatically introducing a defined (in terms of length) nonisotopic label into preexisting oligodeoxyribonucleotides, but does not block the 3'-terminus for further elongation *(9)*.

2. Materials

1. It is recommended that oligonucleotide primers are fully deprotected and purified from free blocking groups by ethanol precipitation from $0.3M$ NaOAc, pH 4.8, with 3 vol of absolute EtOH ($-20°C$ for 60 min), followed by a wash with 0.5 mL 70% EtOH. Their concentration is determined spectrophotometrically at 260 nm using a factor of 30 $\mu g/A_{260}$.
2. Fluorescein-12-riboUTP is a product of Boehringer Mannheim (Mannheim, Germany) (*see* Note 1). It can be stored as a 1-mM solution for several weeks at $-20°C$. It should be kept on ice and shielded from strong light sources during use.
3. Biotin-16-riboUTP is obtained from Boehringer Mannheim.

4. 25 mM CoCl$_2$ and 5X terminal transferase buffer (1M K-cacodylate, 125 mM Tris-HCl, 1.25 mg/mL BSA, pH 6.6) are supplied together with terminal transferase by Boehringer Mannheim (but *see* Note 2). The CoCl$_2$ is diluted to its stock concentration of 5 mM with water and stored frozen. The enzyme is diluted 1:1 with cold 50% glycerol (to give approx 12.5 U/μL) and can be stored in this form at −20°C.
5. Spin columns are made using empty MicroSpin columns (Pharmacia Biotech, Freiburg) (or similar) using 0.7 mL of a thick slurry of Bio-Gel P6 (Bio-Rad, München) (*see* Note 3). These can be stored for up to a month at 4°C. Just prior to their use, they are inserted into 2-mL Eppendorf tubes and spun at 3000 rpm (720g) for 1 min (Eppendorf 5402 centrifuge).

3. Method

1. The incubation mixture (10 μL total vol) consists of:
   a. 1–2 μL oligonucleotide (approx 3 μg).
   b. 2 μL 5X terminal transferase buffer.
   c. 1.5 μL 5 mM CoCl$_2$.
   d. 2 μL 1 mM fluorescein- or biotin-rUTP.
   e. H$_2$O to 9 μL.
2. The reaction is initiated by the addition of 1 μL (10–20 U) terminal transferase.
3. Incubation is at 37°C for 30 min. In the case of fluorescein, shielding from direct light is advisable (*see* Note 4).
4. 10 μL H$_2$O are added to decrease losses during spin chromatography.
5. The total 20 μL are applied to a prespun Bio-Gel P6 column, and the centrifugation is carried at exactly the same settings as during the prespin, i.e., 3000 rpm (720g), 1 min (*see* Note 5).
6. The concentration of the material eluted, now essentially freed from unincorporated nucleotides, may be determined spectrophotometrically and diluted appropriately. Residual amounts of free ribonucleotides, not being substrates for DNA polymerases, do not interfere with subsequent reactions.
7. The primer is now ready for PCR, for conventional, or for cycle sequencing (*see* Note 6).

4. Notes

1. Fluorescein-12-riboUTP and biotin-16-riboUTP are supplied as stock solutions of 10 mM. They should be diluted to 1 mM with H$_2$O.
2. **Important:** Some batches of enzymes have proven to be unsuitable for the purposes of the labeling reaction described here. There appears to be a 5'-exonuclease-like activity in some preparations, causing a population of labeled primers with heterogeneous 5'-termini and giving rise to characteristic sequences, unreadable near the primer, but improving after 150–200 bases *(10)*. Such primers show, in the case of fluorescein labeling, instead of a single band, two or more fluorescent bands on analysis on polyacrylamide/urea gels (12–15% polyacrylamide, depending on the primer length). The bands are resistant to alkali

treatment (which would hydrolyze any polymeric ribo-tails). Should one suspect the appearance of this phenomenon, a change of enzyme batch or supplier (e.g., the products of MBI Fermentas [Vilnius] and of Amersham [Braunschweig] have been found to be suitable, using the buffer supplied by Boehringer) will alleviate the problem.

3. The gel-filtration medium may be pipeted with a commercial pipet using a disposable tip whose orifice has been enlarged by trimming back the tip by approx 5 mm. Do **not** be tempted to use original MicroSpin material; Sephacryl S-400 has been shown to bind single-stranded DNA irreversibly *(11)*, and we have observed a similar effect with Sephacryl S-200.

4. Under these conditions, the extent of labeling is about 70% for fluorescein and almost 100% for biotin *(9)*.

5. An alternative to spin-column chromatography is butanol precipitation. Ethanol precipitation is not recommended, since fluorescein- or biotin-bearing oligonucleotides are, depending on the base composition, often ethanol-soluble. Butanol precipitation is carried out by adding 180 μL *n*-BuOH to the 20-μL vol of the reaction mixture. After vigorous vortexing, the mixture may be centrifuged directly (14,000 rpm, 15,800g) at room temperature. Unincorporated nucleotides may be partially removed from the pelleted material by careful washing with cold (–20°C) 80% EtOH and a 5-min spin at room temperature. The pellet (clearly discernible, despite the small amounts involved, owing to its orange color) is then dried under gentle vacuum and redissolved in H_2O.

6. For fluorescence-based sequencing, requiring a primer stock solution of 20 ng/μL (approx 3 pmol/μL), we assume a 30% loss of material during purification and dilute the eluate to 100 μL with H_2O without further quantification. One microliter is then taken per sequence reaction. For PCR, quantification is advisable. Cycle sequencing is routinely carried out using Sequitherm (Epicentre, Madison) or Thermosequenase (Amersham, Braunschweig), and the nucleotide concentrations and protocols recommended by the manufacturer. We strongly urge users of primers labeled according to this method **not** to include NaOH in their stop solutions *(6,12)*. Even brief alkali treatment of the ribose link between the 3'-terminal nucleotide of the primer and the first base of the extended DNA chain can lead to its cleavage and consequent loss of label.

Note Added in Proof

With the availability of fluorescein-CTP and fluorescein-ATP (DuPont-NEN, Boston, MA), a more flexible choice of primers becomes possible. For effective labeling with these nucleotides the terminal transferase buffer must be replaced by 1 μL dimethylsulfoxide (DMSO) + 1 μL 0.5M potassium phosphate, pH 7.2. All other conditions remain unchanged.

References

1. Murray, V. (1989) Improved double stranded DNA sequencing using the linear polymerase chain reaction. *Nucleic Acids Res.* **17,** 8889.

2. Blakesley. R. W. (1993) Cycle sequencing, in *Methods in Molecular Biology*, vol. 23, *DNA Sequencing Protocols* (Griffin, H. G. and Griffin, A. M., eds.), Humana, Totowa, NJ, pp. 209–217.

3. Smith, L. M., Kaiser, R. J., Sanders, J. Z., and Hood, L. E. (1987) The synthesis and use of fluorescent oligonucleotides in DNA sequence analysis. *Methods Enzymol.* **155**, 260–301.

4. Prober, J. M., Trainor, G. L., Dam, R. J., Hobbs, F. W., Robertson, C. W., Zagursky, R. J., Cocuzza, A. J., Jensen, M. A., and Baumeister, K. (1987) A system for rapid DNA sequencing with fluorescent chain-terminating dideoxy-nucleotides. *Science* **238**, 336–341.

5. Voss, H., Wiemann, S., Wirkner, U., Schwager, C., Zimmermann, J., Stegemann, J., Erfle, H., Hewitt, N. A., Rupp, T., and Ansorge, W. (1992) Automated DNA sequencing system resolving 1000 bases with fluorescein-15-*dATP as internal label. *Methods Mol. Cell Biol.* **3**, 153–155.

6. Ansorge, W., Zimmermann, J., Erfle, H., Hewitt, N., Rupp, T., Schwager, C., Sproat, B., Stegemann, J., and Voss, H. (1993) Sequencing reactions for ALF (EMBL) automated DNA sequencer, in *Methods in Molecular Biology*, vol. 23, *DNA Sequencing Protocols* (Griffin, H. G. and Griffin, A. M., eds.), Humana, Totowa, NJ, pp. 317–356.

7. Roychoudhury, R. and Kössel, H. (1971) Synthetic polynucleotides. Enzymatic synthesis of ribonucleotide terminated oligodeoxynucleotides and their use as primers for the enzymatic synthesis of polydeoxynucleotides. *Eur. J. Biochem.* **22**, 310–320.

8. Kössel, H. and Roychoudhury, R. (1971) Synthetic polynucleotides. The terminal addition of riboadenylic acid to deoxyoligonucleotides by terminal deoxy-nucleotidyl transferase as a tool for the specific labelling of deoxyoligonucleotides at the 3' ends. *Eur. J. Biochem.* **22**, 271–276.

9. Igloi, G. L. and Schiefermayr, E. (1993) Enzymatic addition of fluorescein- or biotin-riboUTP to oligonucleotides results in primers suitable for DNA sequencing and PCR. *BioTechniques* **15**, 486–497.

10. Schiefermayr, E. and Igloi, G. L. (1995) Degradation of DNA sequencing primers by a terminal transferase-associated exonuclease. *Anal. Biochem.* **230**, 180–182.

11. Nehls, M. and Boehm, T. (1993) S-300 matrix irreversibly binds single-stranded nucleic acids. *Trends Genet.* **9**, 336–337.

12. Zimmermann, J., Wiemann, S., Voss, H., Schwager, C., and Ansorge, W. (1994) Improved fluorescent cycle sequencing protocol allows reading nearly 1000 bases. *BioTechniques* **17**, 302–307.

4

Direct Sequencing of Double-Stranded PCR Products with the Sequenase Kit and [α-^{35}S] dATP

Jean-Laurent Casanova

1. Introduction

The polymerase chain reaction (PCR) products are double-stranded linear DNA molecules. Although digestion with a set of restriction enzymes, hybridization with an internal probe, detection of a single-stranded chain polymorphism, or of a site for specific chemical cleavage may all provide useful information on the amplified products, clear-cut identification of nucleic acids is best achieved by sequencing. When a PCR fragment is heterogeneous, cloning in a vector may be required for sequencing each individual molecule independently. In many cases, however, regions of or often the entire PCR product is homogeneous, and direct sequencing without cloning may be undertaken. We have developed a simple and fast method for directly sequencing linear double-stranded DNA molecules, such as PCR products *(1)*, which we describe in detail below.

This method is direct, since it allows the sequencing of a PCR product without an intermediate cloning step and without the generation of the single-stranded linear DNA template by an additional step, such as asymmetrical PCR or separation of a biotinylated strand. This method is also simple, fast, and cheap, since it makes use of the most common sequencing reagents, namely the Sequenase kit (United States Biochemicals, Cleveland, OH) and [α-^{35}S] dATP, without any modification of the primer or of the reaction buffer. The sequencing of double-stranded linear DNA by this method is achieved mainly by optimization of incubation times and/or temperatures.

The major problem in sequencing linear double-stranded DNA is the renaturation of the template, which affects one or several steps of the sequencing reaction, such as the annealing of the primer to the single-stranded template,

From: *Methods in Molecular Biology, Vol. 65: PCR Sequencing Protocols*
Edited by: R. Rapley Humana Press Inc., Totowa, NJ

the labeling of the newly synthesized fragment with [α-^{35}S] dATP, or the extension and termination of the synthesis in the presence of dideoxy nucleotides. The following three conditions have been considered in this method to minimize the renaturation or the effect of renaturation on the sequencing process.

1. Annealing of the primer to the template and minimal renaturation of the template are best achieved by rapid transfer of the sample from +100 to −70°C.
2. A 10- to 100-fold primer-to-template molar excess favors annealing of the primer to the template during rapid cooling.
3. Labeling is optimal at room temperature between 15 and 45 s.

Extension conditions (time and temperature) have less influence on the outcome of the sequencing reaction than those applied to the annealing and labeling steps.

2. Materials
2.1. Specific Materials

1. Sequenase version 2.0 kit (United States Biochemicals), including DTT (0.1M), dNTP mix (5X) for dGTP sequencing, four ddNTP mixtures, enzyme dilution buffer (1X), reaction buffer (5X), Mn^{2+} buffer (1X), and stop solution.
2. [α-^{35}S] dATP (1 mCi/pmol/80 µL) (e.g., Du Pont, Boston, MA).
3. 96-well round-bottom plate and lid (e.g., Dynatech, Chantilly, VA).
4. Geneclean II kit (Bio 101, La Jolla, CA).
5. Linear acrylamide (0.25%, as described in ref. 2) or other carrier (glycogen, tRNA).
6. Metallic racks for Eppendorf tubes; metallic water bath resistant to boiling; dry ice ethanol bath.
7. Salad (lettuce) spinner or 96-well plate centrifuge.

2.2. Optional Materials

1. Centricon 100 microconcentrators (Amicon, Danvers, MA).
2. Qiagen PCR purification kit (Diagen, Düsseldorf, Germany).
3. Electroelutor (International Biotechnologies Inc IBI, New Haven, CT).

2.3. Common Materials

1. Agarose.
2. Ethidium bromide, 10 mg/mL.
3. Urea.
4. Acrylamide:bis-acrylamide (19:1) 40%.
5. Ammonium persulfate 25%.
6. N,N,N',N'-Tetramethylethylenediamine (TEMED).
7. Sodium acetate, 3M (pH 5.2).
8. Ethanol and ethanol:water (80:20).
9. SpeedVac or similar vacuum dryer.

10. Oven.
11. Sequencing gel device.
12. TBE buffer: 10X (54 g Tris base, 27.5 g boric acid, 20 mL 0.5M EDTA ([pH 8.0]/L).
13. DNA loading buffer 6X: Ficoll 15%, bromophenol blue 0.25%, xylene cyanol 0.25%.

3. Method

3.1. Purification of the PCR Product

1. Precipitate the PCR products with 0.1 vol of 3M sodium acetate, pH 5.2, 3 vol of ethanol (carrier if needed, such as 2 µL of linear acrylamide, 0.25%) and freezing at –70°C for 15 min.
2. Centrifuge the sample at 10,000g for 15 min, discard the supernatant, rinse the pellet with 80% ethanol, spin down again at same speed for 3 min, discard the supernatant, and dry the pellet under vacuum for 3 min.
3. Resuspend the pellet in 15 µL of 1X DNA loading buffer, load onto a 2% agarose gel (for a PCR product of 200–800 bp) stained with ethidium bromide (2 µL for 50 mL), and run at 5 V/cm for 1 h in 1X TBE buffer (*see* Note 2).
4. Cut out the band(s) of interest under UV transillumination and store it in a labeled Eppendorf tube(s) at 4°C. Elute DNA from gel slice. Excellent results are often obtained with the Geneclean II kit, but other methods (electroelution, and so forth) may be used. Resuspend the PCR product in 10 µL of water.

3.2. Sequencing of the PCR Product

1. Set up a dry ice–ethanol bath, a boiling water bath, and several ice boxes. Thaw out the reagents of the Sequenase version 2.0 kit (not the enzyme). Label the lid of the 96-well plate with the names of the templates at the emplacement corresponding to the last four wells of each line.
2. For eight PCR products to be sequenced, prepare eight Eppendorf tubes each containing 6 µL of water, 2 µL of 5X reaction buffer, 1 µL of primer (10 µM), and 2 µL of template (ideally around 0.5 pmol) (*see* Notes 3-5).
3. The volumes indicated in this section are for eight sequencing reactions. Prepare in an Eppendorf tube a sequencing mix containing 8 µL DTT (0.1M), 16 µL 1X dNTP mix for dGTP sequencing, and 4 µL [α-^{35}S] dATP. In a second Eppendorf, prepare 14 µL of 1X enzyme dilution buffer. Keep both tubes on ice.
4. Put 2.5 µL of each ddNTP mixture in the last four wells of each line of the plate, for example, in the following order: ddGTP, ddATP, ddTTP, and ddCTP.
5. Place the eight Eppendorf tubes in a rack that maintains the lids firmly closed and incubate the rack in the closed boiling water bath for 3–5 min (*see* Note 6).
6. Add 8 µL of 1X Mn^{2+} buffer to the sequencing mix while the samples are being boiled.
7. Transfer the rack from the boiling bath to ice, and then transfer the Eppendorf tubes immediately to the dry ice–ethanol bath (*see* Note 7).
8. Add 2 µL of Sequenase to the 14-µL enzyme dilution buffer, transfer this diluted Sequenase to the sequencing mix, and homogenize by pipeting. Keep the tube on ice.

9. Take the first tube from the dry ice–ethanol bath, wipe off the alcohol from the external walls of the tube with paper towel, turn the tube upside down when opening the lid to prevent anything from falling into the tube, and add 5.5 µL of the sequencing mix on the top of the frozen pellet.

10. Warm up the tube in your hand, and help to thaw and mix the sample by rotating the pipet tip.

11. When the sample is melted, i.e., as soon as you cannot see any ice particles, incubate the tube at room temperature for 15–45 s.

12. Transfer 3.5 µL to the left border of each of the four wells of a line containing the ddNTP mixtures, cover the plate with the lid, and spin down briefly in the salad spinner (*see* Note 8).

13. Incubate the plate at room temperature (or at 37°C) until step 15.

14. Repeat the procedure for each of the remaining seven tubes (*see* Note 9).

15. Immediately add 4 µL of stop solution to the right border of each of the 32 wells, and spin down.

16. Heat the plate in an oven at 80°C for 2 min, load 4 µL of the sequencing reaction on $8M$ urea, 6% acrylamide sequencing gel, run at 50°C in 1X TBE buffer, dry the gel after elution of urea, and expose overnight (*see* Notes 10–16) (*see* Chapter 1).

4. Notes

4.1. Technical Comments

1. Prepare a 5% acrylamide solution (without bis-acrylamide) in 40 mM Tris-HCl, 20 mM sodium acetate, and 1 mM EDTA, pH 7.8. Add 1/100 vol of 10% ammonium persulfate and 1/1000 vol of TEMED, and let polymerize for 30 min. When the solution has become viscous, precipitate the polymer with 2.5 vol of ethanol, centrifuge, and redissolve the pellet in 20 vol of water by shaking overnight. The 0.25% linear polyacrylamide solution can be stored in the refrigerator for several years.

2. The purification of the PCR products is also possible with Centricon-100 filters or Qiagen columns. These procedures separate the primers from all PCR products. Therefore, they may be applied only when a single product is generated by PCR, whereas the agarose gel electrophoresis allows the independent purification and subsequent sequencing of several PCR products from the same reaction. In addition, the presence of a smear or even of infravisible PCR products may alter the sequencing reaction. Finally, the ultimate cloning of these semipurified molecules, if necessary, is partly compromised. Altogether, it is recommended that as a first choice purifying PCR products on an agarose gel and comparing (for each particular type of PCR) the other methods to this reference purification.

3. The amount of template should be ideally around 0.5 pmol (as determined by agarose gel quantification). If lower, load the entire sequencing reaction on gel.

4. The amount of sequencing primer is dependent on the amount of template and should stay between a 10- and 100-fold molar excess.

5. The amount of linear acrylamide per sample should stay below 1 µg in the course of the sequencing process. This information is not known for the other carriers, such as glycogen or tRNA.

6. Do not heat the samples on a solid heating block or the sample will evaporate. Covering the boiling water bath creates uniform temperature conditions, preventing any evaporation of the samples.

7. To prevent renaturation of the template, freeze the samples as rapidly as possible, that is, in a dry ice ethanol–bath, and not in dry ice or in a freezer.

8. The salad (lettuce) spinner has three major advantages over a classical centrifuge: It vortexes the samples, it is much faster to start and to stop, and it is cheaper. If you do not find a salad spinner, use a classical centrifuge, and spin down the plates as briefly as possible. Alternatively, tap the plate vigorously on the bench top to bring the drops to the bottoms of the wells.

9. The protocol described above can be adapted for handling 1–12 tubes and possibly more. Indeed, the extension time and temperature appear to be of little importance, at least in the reading of the first 150 bp.

4.2. General Comments

10. The quality of the sequence depends primarily on the purity and homogeneity of the PCR product. If your sequence is poorly readable, first investigate the PCR conditions. Check that you wish to sequence the specific product of a single gene from a single cell type and that your primers are specific for that gene only. However, heterogeneous PCR products may be deliberately sequenced by this method once the technique is set up, and it works well in routines with homogeneous products (*see* Note 15).

11. The quality of the sequence also depends on the location of the primer. The same primers used to generate the PCR product usually work well, but nested primers often work better. Given the very short labeling time, it is also very important to check that the [α-^{35}S] dATP can be incorporated within the next 10 nucleotides following the primer. This is an important point to consider in the design of the primer. The more incorporated dATPs in the DNA strand synthesized by Sequenase during the labeling phase, the better. Alternatively, other labeled nucleotides may be used.

12. Some combinations of primer-PCR product may not work under these conditions, possibly because of the faster self-reannealing of the template. You may try to change the primer. If it still does not work, you may try to label the primer with ^{32}P (to avoid the labeling step), and otherwise apply the same protocol. Alternatively, generation of the single-stranded linear template by asymmetrical PCR or by separation of a biotinylated strand may help. In some cases, however, probably because of a strong secondary structure of the single-stranded template or of the primer, direct sequencing with Sequenase cannot be achieved. Direct sequencing at higher temperatures with thermostable polymerases may help to overcome the difficulty. Occasionally, molecular cloning will be required for sequencing.

13. Products ranging in size from 200 to 800 bp have been sequenced with this method. Above or below may be possible, but has not been attempted. The length of the readable sequence is usually at least 100 bp and at most 200 bp. Longer sequences are not obtained as a rule, and other methods should be used for such purposes.

14. Because PCR also amplifies mishybridizations of the primers, the expected size of a PCR product is only a good indication of specificity, but not as good as in

classical nonamplified hybridizations. Thus, digestion of the PCR product with restriction enzymes, hybridization with an internal probe, research of a single-stranded chain polymorphism, of a specific chemical cleavage, or ideally sequencing of all or part of the product should be performed in all cases. This method allows rapid and simple identification of any PCR product for that purpose.

15. This procedure is also the method of choice when you want sequence information that does not extend over 200 bp. A very good application is, for example, the sequencing of T-cell receptor (TCR) junctional regions amplified from T-cell clones *(2–5)* or single T-cells *(6)*. Note that a PCR product that is heterogeneous but shows regions of homology or identity, such as a TCR PCR product derived from a polyclonal T-cell population, can also be directly sequenced *(3,4,7)*.

16. Because there is no cloning step, most thermostable polymerase misincorporations are not detected by the sequencing of this polyclonal product. Sequencing several independent PCR products is therefore not necessary in most cases.

References

1. Casanova, J.-L., Pannetier, C., Jaulin, C., and Kourilsky, P. (1990) Optimal conditions for directly sequencing double-stranded PCR products with sequenase. *Nucleic Acids Res.* **18,** 4028.
2. Casanova, J.-L., Romero, P., Widmann, C., Kourilsky, P., and Maryanski, J. L. (1991) T cell receptor genes in a series of class I Major Histocompatibility Complex restricted cytotoxic T lymphocyte clones specific for a *Plasmodium berghei* nonapeptide: implications for T cell allelic exclusion and antigen-specific repertoire. *J. Exp. Med.* **174,** 1371–1383.
3. Casanova, J.-L., Cerottini, J.-C., Matthes, M., Necker, A., Gournier, H., Barra, C., Widmann, C., MacDonald, H. R., Lemonnier, F., Malissen, B., and Maryanski, J. L. (1992) H-2 restricted cytolytic T lymphocytes specific for HLA display T cell receptors of limited diversity. *J. Exp. Med.* **176,** 439–447.
4. Casanova, J.-L., Martinon, F., Gournier, H., Barra, C., Regnault, A., Pannetier, C., Kourilsky, P., Cerottini, J.-C., and Maryanski, J. L. (1993) T cell receptor selection by and recognition of two class I MHC restricted antigenic peptides that differ at a single position. *J. Exp. Med.* **177,** 811–820.
5. Romero, P., Casanova, J.-L., Cerottini, J.-C., Maryanski, J. L., and Luescher, I. (1993) Differential TCR photoaffinity labeling among H-2K^d restricted CTL clones specific for a photoreactive peptide derivative. Labeling of the α chain correlates with Jα segment usage. *J. Exp. Med.* **177,** 1247–1256.
6. Maryanski, J. L., Jongeneel, C.-V., Bucher, P., Casanova, J.-L., and Walker, P. R. (1996) Single-cell PCR analysis of TCR repertoire selected by antigen in vivo: a high magnitude CD8 response is comprised of very few clones. *Immunity* **4,** 47–55.
7. MacDonald, H. R., Casanova, J.-L., Maryanski, J. L., and Cerottini, J.-C. (1993) Oligoclonal expansion of MHC class I restricted CTL during a primary response in vivo: direct monitoring by flow cytometry and polymerase chain reaction. *J. Exp. Med.* **177,** 1487–1492.

5

Direct Sequencing by Thermal Asymmetric PCR

Georges-Raoul Mazars and Charles Theillet

1. Introduction

Direct sequencing of PCR products *(1)* has proven to be a powerful method in the generation of nucleic acid sequence data. Using these techniques, it is possible to produce microgram quantities of pure target DNA and subsequently its nucleotide sequence in a few hours, even, theoretically, from one single RNA or DNA molecule. However, problems have been encountered, and these have been attributed to the strong tendency of the short double-strand DNA templates to reanneal. In fact, compared to double-stranded plasmid DNA, which can be permanently denatured by alkali treatment and then form intermolecular interactions compatible with good sequencing efficiency, optimized conditions for direct sequencing are required before reannealing with short PCR product.

In order to obviate this, strategies have been developed, such as the generation of single-strand DNA template by asymmetric PCR *(2,3)*. Methods use either a disequilibrated concentration ratio between the two primers or a two-step amplification. Both have their shortfalls. The first method is based on a large number of cycles, which is a potential source of misincorporation of errors, and optimized conditions enough to produce single-strand DNA that are strongly primer-dependent. Moreover, it often has been the case that only one strand can easily be sequenced. The second case requires two physical separation steps, in which product contamination may occur.

Here we propose a method combining the advantages of both symmetric and asymmetric PCR. It is based on a thermal asymmetry between the T_m of both primers. Annealing temperature of each primer is calculated with the formula: $69.3 + 0.41$ (%GC) $- 650/L$ (L = primer length). PCR primers are designed in order to obtain a difference in T_m of at least 10°C. In the first step,

From: *Methods in Molecular Biology, Vol. 65: PCR Sequencing Protocols*
Edited by: R. Rapley Humana Press Inc., Totowa, NJ

double-stranded material is produced during 20–25 cycles (to minimize the yield of spurious products) using the lower T_m. During the second step, single-stranded DNA is generated using the higher T_m (Fig. 1). Consequently, one primer is dropped out and linear amplification is obtained. The final quantity of single-stranded product is comparable with the one produced by Gyllensten and Erlich's method *(2)*. We applied thermal asymmetry to several sequences, which, in our hands, were difficult to sequence both from double-stranded DNA or with the Gyllensten and Erlich asymmetric PCR products (Fig. 2). These PCR fragments comprised: (1) exons 7 and 8 of the p53 gene and (2) exon 1 of HRAS. This latter sequence is particularly GC-rich, and as part of another study, primer A was synthesized with a 40-base long GC stretch (in order to make a GC clamp). In conclusion, thermal asymmetric PCR allows direct sequencing of both strands with high reproducibility and reduced risk of contamination.

2. Materials

All solutions should be made according to the standard required for molecular biology, such as molecular-biology-grade reagents and sterile distilled water. All reagents for sequencing are available commercially.

2.1. PCR Amplification

1. Primers synthesized on Applied Amplifications, and PCR was performed on a Perkin-Elmer Cetus thermal cycler.
2. 1X *Taq* polymerase buffer: 10 mM Tris-HCl, pH 8.3, 2 mM MgCl$_2$, 50 mM KCl, 0.01% gelatin, 100 µM of dNTPs.
3. *Taq* polymerase was purchased from Perkin-Elmer and used at 1 U/reaction.

2.2. Purification and Sequencing of the PCR Product

2.2.1. Sequencing Reagents

1. Annealing buffer (5X concentrate): 200 mM Tris-HCl, pH 7.5, 100 mM MgCl$_2$, 250 mM NaCl, 0.1M Dithiothreitol (DTT).
2. Labeling nucleotide mixture (one for each dideoxy nucleotide): Each mixture contains 80 µM dGTP, 80 µM dATP, 80 µM dTTP, 80 µM dCTP, and 50 mM NaCl. In addition, the "G" mixture contains 8 µM dideoxy-dGTP; the "A" mixture, 8 µM ddATP; the "T," 8 µM ddTTP; and the "C" 8 µM ddCTP.
3. Stop solution: 95% formamide, 20 mM EDTA, 0.05% bromophenol blue, and 0.05% xylene cyanol FF.
4. Labeled dATP is [α-^{35}S] dATP from Amersham, and specific activity should be 1000–1500 Ci/mol.

2.2.2. Purification

Purification should be performed in Centricon 30 column (Amicon).

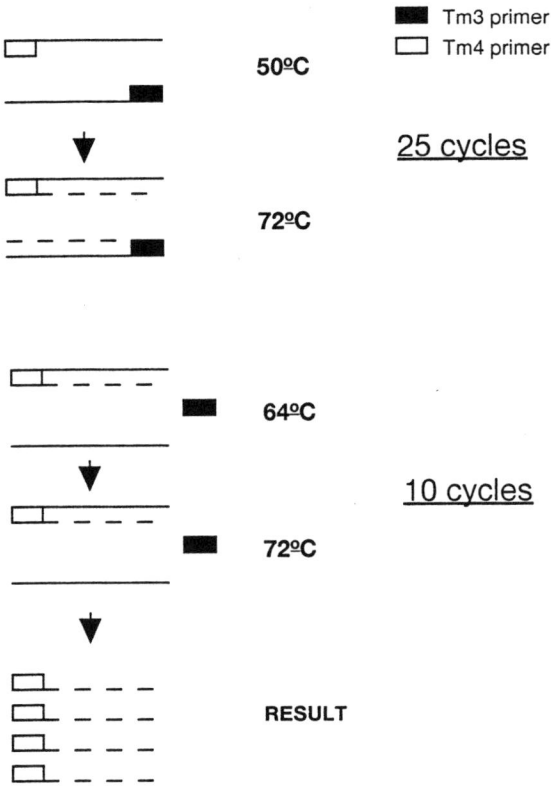

Fig. 1. Schematic representation of thermal asymmetric PCR.

3. Methods

3.1. PCR Conditions

PCR conditions should be as follow: in a total volume of 25 µL, 20 pmol of each primer, 1X *Taq* polymerase buffer, and 1 U of *Taq* polymerase were incubated with 50 ng of genomic DNA. P53 primer A1 cttagtacctgaagggtgaaatattc ($T_m1 = 60°C$), P53 primer B1 gtagtggtaatctactgggacggaacagc ($T_m2 = 69°C$), P53 primer A2 taatctactgggacgga ($T_m3 = 50°C$), P53 primer B2 cccaagacttagtacctgaagggtg ($T_m4 = 64°C$). Cycling conditions were for 25 cycles: 92°C (30 s); T_m1 or 3 (30 s); 72°C (90 s), followed by 10 cycles: 92°C (30 s); T_m2 or 4 (30 s); 72°C (90 s).

3.2. Preparation of the PCR Product

1. Transfer PCR product directly by pipeting in a Centricon 30, and add 2 mL of water.
2. Spin at 5000*g* in a fixed-angled rotor in a Beckman-type centrifuge for 30 min at room temperature.

A C G T

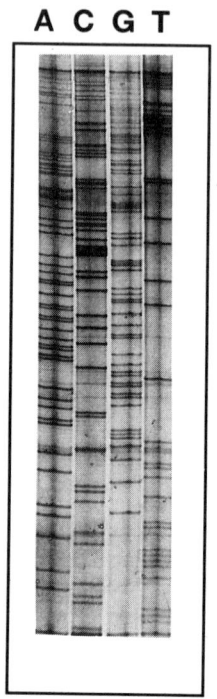

Fig. 2. Autoradiograph of a PCR product sequenced by thermal asymmetric PCR.

3. Add 2 mL of water. Spin again for 30 min. Invert column and spin for 5 min at 1500g. This procedure efficiently removes the excess of dNTPs from the PCR reaction. Volume recovered is typically 20–50 μL.
4. Typically, 7 μL of this purified product are used for single-strand sequencing according to the manufacturer's directions of United States Biochemicals (Cleveland, OH).

3.3. Sequencing Protocol

3.3.1. Annealing Template and Primer

1. For each template, a single annealing (and subsequent labeling) reaction is used. Combine the following:
   a. Primer   0.5 pmol (1 μL)
   b. DNA   7 μL
   c. Annealing buffer   2 μL
2. Warm the capped tube to 65°C for 2 min, and then allow the mixture to cool slowly to room temperature over a period of about 30 min.

3.3.2. Labeling Reaction

1. To the annealed template-primer add the following:

a.	DTT (0.1M)	1 µL
b.	Labeling nucleotide mix	2 µL
c.	[α-^{35}S] dATP	5 µCi (typically 0.5 µL)
d.	Sequenase	3 U from United States Biochemicals

2. Total volume should be approx 15 µL; mix thoroughly and incubate for 2–5 min at room temperature.

3.3.3. Termination Reactions

1. Label four tubes "A," "C," "G," and "T." Fill each with 2.5 µL of the appropriate dideoxy termination mixture.
2. When the labeling reaction is complete, transfer 3.5 µL of it to the tube (prewarmed to 37°C) labeled "G." Similarly, transfer 3.5 µL of the labeling reaction to each of the other three tubes ("A," "T," and "C").
3. After 2–5 min of incubation at 37°C, add 4 µL of stop solution to each termination reaction, mix, and store on ice.
4. To load the gel, heat the samples to 75–80°C for 2 min, and load 2–3 µL in each lane. Prerun a sequencing gel for 30 min, load, and run until bromophenol is just out of the gel.
5. Fix gel as usual and dry on Whatman 3MM paper. Correct sequencing yields a detectable signal using a bench-top Geiger counter.
6. Expose overnight without Saran™ paper at room temperature.

4. Notes

1. Instability of diluted solution of primers conserved at –20°C can sometimes be problematic. We recommend storing oligonucleotides as a dried powder and resuspending them in water prior to use.
2. Estimation of the yield of single-strand DNA produced can be achieved by Southern blotting: run a 2% agarose gel, blot following standard conditions, and probe with one of the PCR primers: two bands should appear if you use higher T_m primer or only one band if you use lower T_m primer (corresponding to double-stranded DNA). An alternative strategy is to add α-dCTP[^{32}P] for the second step of amplification at high temperature: labeled single-stranded DNA should be exclusively produced.
3. We also sequenced PCR fragments following SSCP analysis. In this case, shifted SSCP bands were excised from the gel with a sterile razor blade and eluted in 50 µL of distilled water for 1 h at 65°C. A 1.5-µL aliquot of the eluate was subjected to thermal asymmetric PCR.
4. A good sequence can be obtained even if no primer is added for the sequencing reaction. The reason is that the low T_m primer from PCR is not completely removed by Centricon 30 purification.
5. The present protocol has been optimized for classical radioactivity labeled DNA sequencing, but should easily be adapted to automated fluorescent sequencing using fluorescent dye terminators.

References

1. Saiki, R. K., Gelfand, D. H., Stoffel, S., Scharf, S. F., Higuchi, R., Horn, R. T., Mullis, K. B., and Erlich, H. A. (1988) Primer-directed enzymatic amplification of DNA with a thermostable DNA polymerase. *Science* **239,** 487–491.
2. Gyllensten, U. B. and Erlich, H. A. (1988) Generation of single-stranded DNA by the polymerase chain reaction and its application to direct sequencing of the HLA-DQA locus. *Proc. Natl. Acad. Sci. USA* **85,** 7652–7656.
3. Wilson, R. K., Chen, C., and Hood, L. (1990) Optimization of asymmetric polymerase chain reaction for rapid fluorescent DNA sequencing. *BioTechniques* **8,** 184–189.

6

Rapid Sequencing of cDNA Clones

Direct Sequencing Using Sequential Linear/Asymmetric PCR

Ivor J. Mason

1. Introduction

Following the isolation of a clonal recombinant phage or plasmid after screening a cDNA library, the first analyses that are routinely performed are the determination of the insert size and the sequence of the 3'- and 5'-ends of the cloned fragment. Prior to the use of PCR, this was a laborious process, especially when the cDNA library had been constructed using a vector derivative of bacteriophage λ. I recently described a PCR-based protocol that rapidly generated this information from single-phage plaques and, additionally, produced material suitable for subcloning into plasmid vectors *(1)*. The procedure, which is outlined in Fig. 1, involves an initial PCR reaction using an oligonucleotide primer pair which flank the cDNA insertion site to produce sufficient quantities of double-stranded DNA for sequencing. An aliquot of this reaction is then subjected to further PCR, but with only one of the primers. Any residual second primer is rapidly depleted during this second reaction and large quantities of single-stranded DNA are generated. This material is then used as a template for sequencing using the second oligonucleotide to prime the reaction.

The use of linear PCR alone followed by direct sequencing of the double-stranded product could potentially be employed to analyze phage inserts. However, in my experience, this approach does not consistently generate data of acceptable quality. This is probably because, in part, of the presence of short PCR products, which are the result of incomplete DNA synthesis during the amplification procedure. The second, asymmetric PCR overcomes these problems, since only full-length products of this reaction can anneal the second (sequencing) primer. In addition, the use of single-stranded DNA as a template generally allows each gel to be read further.

From: *Methods in Molecular Biology, Vol. 65: PCR Sequencing Protocols*
Edited by: R. Rapley Humana Press Inc., Totowa, NJ

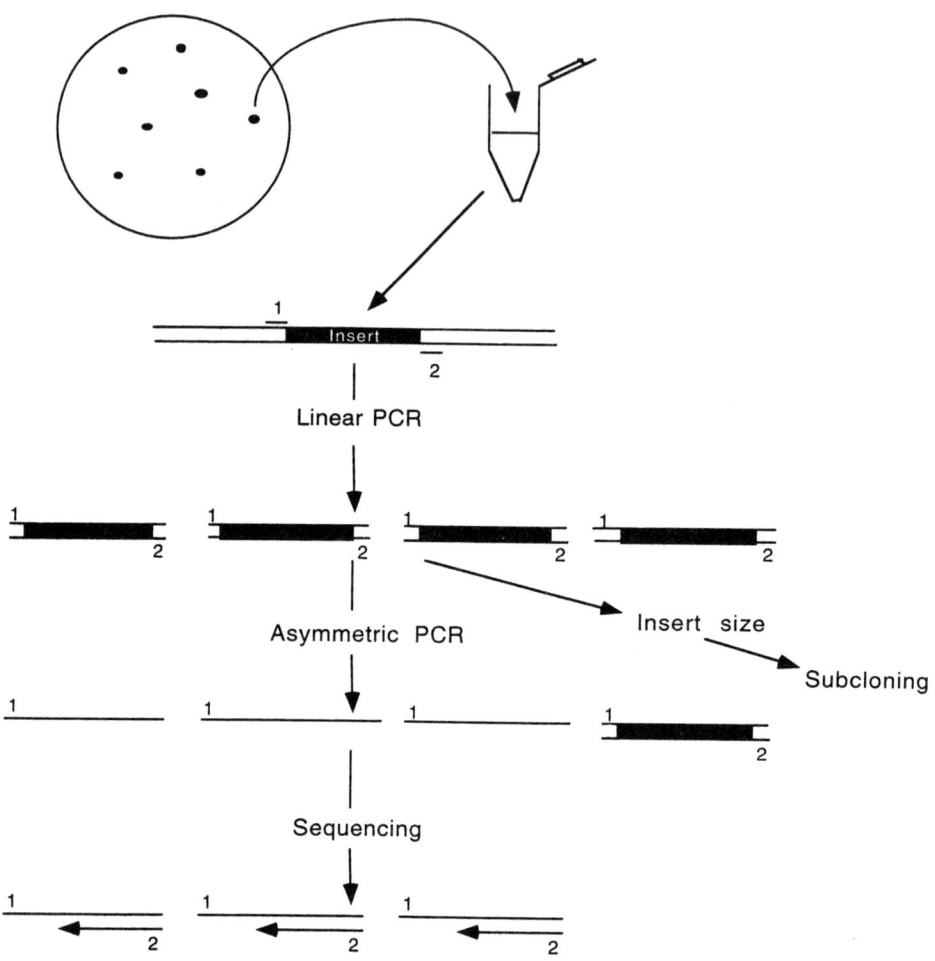

Fig. 1. Schematic diagram of the protocol for sequencing from single-phage plaques.

2. Materials

All solutions should be made to the standard required for molecular biology using molecular-biology-grade reagents and sterile distilled water.

1. 2X SM buffer: $0.2M$ NaCl, 16 mM MgSO$_4$, 40 mM Tris-HCl, pH 7.5, 4% (w/v) gelatin, autoclaved.
2. 10X PCR buffer: 100 mM Tris-HCl, pH 8.4, $0.5M$ KCl, 0.01% (w/v) gelatin.
3. Deoxynucleotide triphosphates: diluted from stock solutions (Boehringer Mannheim, Lewes, UK).
4. *Taq* polymerase: either AmpliTaq™ (Perkin Elmer Cetus) or *Taq* polymerase (Promega, Madison, WI).

5. DNA primers: suitable primers can be purchased from several companies that specialize in molecular biology products.
6. Mineral oil (Merck, Southampton, UK).
7. TE: 10 mM Tris-HCl, pH 8.0, 1 mM EDTA.
8. DNA sequencing kits: available from several manufacturers. We routinely use Sequenase™ (Amersham International, Slough, UK) and TaqTrack (Promega, Madison, WI).
9. SeaPlaque™ agarose (FMC Bioproducts, Rockland, MD).

3. Methods

3.1. Amplification of the Phage cDNA Insert

The following procedure produces sufficient quantities of double-stranded DNA for several sequencing reactions, allows analysis of the phage insert size by gel electrophoresis, and provides material to facilitate subcloning of the insert into more useful vector systems.

1. Core a single bacteriophage plaque from a plate with a sterile capillary or Pasteur pipet, and transfer it to a microfuge tube containing 400 µL of sterile distilled H$_2$O. Allow the phage particles to elute from the agar for 1 h (*see* Note 1).
2. Amplify the DNA insert from the bacteriophage vector using an appropriate primer combination (Table 1). Assemble the reaction as follows in a tube appropriate for the thermal cycler to be used (usually a 0.5-mL microfuge tube):
   a. Distilled water   to 100 µL final volume
   b. 10X PCR buffer   10 µL
   c. 0.25M MgCl$_2$ 4 µL
   d. 20 mM dNTPs   1 µL
   e. Phage eluate   20 µL
   f. Primer 1   2 µL (17 pmol)
   g. Primer 2   2 µL (17 pmol)
   h. *Taq* polymerase   2 U
3. Overlay the reaction with 50 µL of light mineral oil.
4. Amplify with the following conditions:
   One cycle of 94°C for 2 min.
   30 cycles of 94°C for 30 s; 55°C for 30 s; 72°C for 4 min.
   One cycle of 72°C for 10 min.
5. Examine 10 µL of the reaction product by conventional agarose gel electrophoresis. This provides information concerning the size of the cDNA insert, suitable material for subcloning, and so forth.

3.2. Asymmetric Amplification and Sequencing

An asymmetric amplification is performed on an aliquot of the products of the reaction in Section 3.1., and the single-stranded DNA products of this second synthesis are purified and sequenced (*see* Notes 2–5).

Table 1
Sequences of Some Useful Primer Pairs
for Amplifying and Sequencing Using Commonly Encountered Vectors

Vectors	Primer 1	Primer 2
λgt11	5'-GGTGGCGACGACTCCTGGAGCCCG	5'-TTGACACCAGACCAACTGGTAATG
λgt10	5'-CTTTTGAGCAAGTTCAGCCTGGTTAAG	5'-GAGGTGGCTTATGAGTATTTCTTCCAGGG
λZAP and pBluescript	5'-TAATACGACTCACTATAGGG	5'-ATTAACCCTCACTAAAGGGA
pGEM	5'-TAATACGACTCACTATAGGG	5'-GATTTAGGTGACACTATAG
pUC/M13	5'-GTTTTCCCAGTCACGAC	5'-CAGGAAACAGCTATGAC

1. Take 3 µL of the first reaction products (typically 150–750 fmol of double-stranded template DNA; 5–25 ng of a 0.5-kb amplification product) and amplify using 17 pmol of one of the original primers. The second primer is omitted from this reaction, but otherwise amplification conditions are as in Section 3.1., step 4.
2. Transfer the PCR products of this reaction to a fresh microfuge tube with as little of the mineral oil as possible, add 100 µL of 4*M* ammonium acetate, and 400 µL of absolute ethanol.
3. Vortex the mixture, and incubate on ice for 15 min. Centrifuge at 12,000*g* for 10 min. This process removes most of the unincorporated nucleotides and primer.
4. Wash the pellet briefly with 100 µL 70% (v/v) ethanol, air-dry, and resuspend in 20 µL TE.
5. Sequence this single-stranded DNA using a commercially available DNA sequencing kit (Sequenase™, Amersham International, Slough, UK). Sequence half of the resuspended DNA using 840 fmol of the second primer from the original PCR reaction using the manufacturer's protocol for single-stranded DNA template. Analyze the products by standard denaturing polyacrylamide gel electrophoresis (*see* Notes 6 and 7 *[2]*) (*see* Chapter 1).

4. Notes

1. The phage particles can be used for PCR for several months if stored at 4°C. However, the titer of viable phage drops considerably on storage. Therefore, it is recommended to transfer 50 µL of the eluate to an equal volume of 2X SM buffer immediately. Store over 10 µL chloroform at 4°C for a high-titer phage stock.
2. Sequencing large cDNA inserts: The conditions described above have proven suitable for obtaining sequence information from inserts of up to 1.5 kb. However, the conditions for both PCR reactions must be adjusted for larger inserts. This can be reliably achieved by the use of the Taq Extender™ PCR additive (Stratagene, La Jolla, CA) in combination with the manufacturer's buffer system. The concentrations of both the templates and primers are not altered, and neither are the times for denaturation (94°C) or annealing of primers (55°C). However, the manufacturer's guidelines for the extension times (72°C) are followed.
3. Sequencing GC-rich regions: Regions of template that are rich in G and C nucleotides frequently cause premature termination of standard sequencing reactions, which appears as bands at the same position in all four lanes on the autoradiograph of the sequencing gel. A simple and reliable way to overcome this problem is to sequence the second aliquot of single-stranded DNA using a thermophilic DNA polymerase. The TaqTrack™ sequencing system (Promega) has proven effective in this regard.
4. Sequencing M13 and other phage with a single-stranded DNA genome can be performed on single plaques using the above protocol.
5. Sequencing plasmid DNA: This can be performed using the above procedure, with the exception that a single colony is picked with a sterile tip, touched onto

an agar plate to provide a stock of bacteria, and the tip is then agitated in the first PCR reaction mix assembled as below:

a. Distilled water   to 100 μL final volume
b. 10X PCR buffer   10 μL
c. 0.25*M* MgCl$_2$ 4 μL
d. 20 m*M* dNTPs   1 μL
e. Primer 1   2 μL (17 pmol)
f. Primer 2   2 μL (17 pmol)
g. *Taq* polymerase   2 U

The rest of the procedure is as described above, and this will generate acceptable data. However, the following protocol provides a more rapid alternative procedure:

a. A single colony is picked with a sterile tip, touched onto an agar plate, and grown to provide a stock of bacteria. The same tip is then introduced into the PCR reaction mixture as above, and amplification is performed as described in Section 3.1., step 4.
b. The reaction products are examined on a 1% (w/v) low-melting-temperature SeaPlaque agarose gel run in TBE buffer as described *(2)*.
c. Fragments to be sequenced are excised with a razor blade in a minimum volume of agarose, and TE is added to a final volume of 300 μL. The gel is melted at 70°C for 15 min.
d. The DNA is purified by standard procedures *(2)*. Briefly, the aqueous phase is extracted with an equal volume of water-saturated phenol and transferred to a fresh tube. The phenol is then back-extracted with a further 100 μL of TE, and the two aqueous phases are combined and extracted once with phenol, once with phenol:chloroform (1:1), and twice with chloroform. In each case, extraction is performed with an equal volume of the organic reagent. The aqueous phase is transferred to a fresh tube, and the DNA is precipitated by the addition of 44 μL 3*M* sodium acetate, pH 5.2, and 1 mL ethanol followed by incubation at 4°C for 15 min and centrifugation at 12,000*g* for 10 min.
e. The pellet is washed with 70% (v/v) ethanol, dried, and resuspended in 14 μL of distilled water.
f. Half of the purified product is then incubated with 10 pmol of one of the two PCR primers at 98°C for 5 min, snap-frozen in liquid nitrogen, allowed to thaw, and then incubated at 65°C for 2 min. It is then sequenced using the Sequenase™ kit and the procedure for double-stranded templates.

6. Sequencing close to the primer: To obtain sequence of the region as close as 22 bases from the primer, add 1 μL each of 0.15*M* sodium citrate and 0.1*M* MnCl$_2$ to the annealing step of the sequencing reaction.

7. Troubleshooting sequencing gels:
    a. No sequence in the body of the gel, but a high-mol-wt band at the top: This is most likely owing to problems with the termination mixes of the sequencing reaction, and they should be replaced.
    b. No signal from the gel at all: Assuming that this is not the result of an error on the part of the experimenter, the most likely causes are:

i. Failure of the second PCR synthesis: check an aliquot on an agarose gel.
ii. Inactivation of the DNA polymerase: try another batch.
iii. Problems with the batch of primer: try a control reaction with the primer/template supplied with the kit, and check the primer sequence if not a commercial preparation.
c. Faint signals are generally owing to too little template; perform another 30 cycles of PCR on the product of Section 3.2.
d. Sequence becomes very faint a short distance from the bottom of the gel: This is generally because of the termination mixes being too concentrated; try diluting them by 50%.
e. Signals at the same level in all four lanes are generally owing to GC-rich regions. *See* Note 3 above.

References

1. Mason, I. J. (1992) Rapid and direct sequencing of DNA from bacteriophage plaques using sequential linear and asymmetric PCR. *BioTechniques* **12,** 60–61.
2. Sambrook, J., Fritsch, E. F., and Maniatis, T. (1989) Molecular Cloning: A Laboratory Manual, 2nd ed. Cold Spring Harbor Laboratory, Cold Spring Harbor, NY.

7

Direct Sequencing of PCR Products Using Chemiluminescent Detection

Bentley A. Atchison and Andrea M. Douglas

1. Introduction

The following protocol describes a method that can be used for the direct sequencing of polymerase chain reaction (PCR) products using nonradioactive detection procedures. It is based on our experience with a method we have previously published *(1)*. Direct sequencing of PCR products is generally more difficult than sequencing single-stranded DNA *(2)* owing to the presence of primers and deoxynucleoside triphosphates (dNTPs) carried over from the PCR into the sequencing phase and by the rapid reannealing of the double-stranded DNA template. These problems are addressed in our method in the following manner:

1. Excess of primers. The same primers used in the PCR are used in the sequencing of the DNA. Specific sequencing products are then detected with biotinylated probes using multiplexing and chemiluminescent procedures. This overcomes the need to use PCR products free of primers *(3)* or complicated procedures such as those using nested primers *(4)*.
2. Excess of dNTPs. Excess dNTPs carried over from the PCR into the sequencing reaction alter the dideoxynucleoside triphosphate (ddNTPs) to dNTP ratio and hence the optimum conditions for sequencing PCR products *(5)*. This can be overcome by using limited amounts of dNTPs in the PCR or, if necessary, a simple gel purification procedure. The latter procedure is necessary if the PCR is relatively inefficient at low dNTP concentrations or is difficult to optimize to produce a specific product.
3. Reannealing of double-stranded DNA templates. Reannealing of template DNA reduces the efficiency of sequencing of DNA products. This is overcome by cycle sequencing using *Taq* polymerase.

From: *Methods in Molecular Biology, Vol. 65: PCR Sequencing Protocols*
Edited by: R. Rapley Humana Press Inc., Totowa, NJ

The method involves four main steps:

1. The PCR amplification of a specific segment of DNA under conditions that minimize the amplification of spurious products.
2. The PCR product is divided into ddNTP termination mixes according to the normal Sanger method and taken through cycle sequencing. The sequencing products are then denatured and electrophoresed through a denaturing polyacrylamide gel. In order that the efficiency of the method may be improved a number of PCR products can be present in any one cycle sequencing reaction and loaded onto the same lane of the sequencing gel.
3. After electrophoresis, the sequencing products are transferred to a nylon membrane by blotting under reduced pressure.
4. The sequencing ladders are probed with biotinylated oligonucleotides specific for each of the DNA strands. As ladders are produced in the sequencing reactions from copying both DNA strands of the PCR product, the sequence of a PCR product can be determined in both directions by the use of appropriate probes. The biotinylated probes are detected by chemiluminescent procedures. As many as six sequences have been detected on the one membrane by alternate stripping/probing of the DNA.

The requirements for the PCR will vary with the PCR product to be amplified. The following procedure has been used to sequence the control region of human mitochondrial DNA and utilizes a gel purification step because of the relatively inefficient amplification of the DNA with 20 μM dNTPs in the PCR. Spurious PCR products can also interfere with the reading of the DNA sequence; gel purification removes these products from the fragment of interest. Other PCR products can be efficiently sequenced without the gel purification step *(1)* but this step is included for completeness should it be required.

2. Materials

All chemicals are analytical grade.

2.1. PCR Procedures

1. Primers/probes: (*see* Note 1).
   Primer: 5'-ccaccattagcacccaaagc-3'
   Primer: 5'-ggagagtagcactcttgtgc-3'
   Probe 1: 5'-ggtacccaaatctgcttccc-3'
   Probe 2: 5'-ccaccatcctccgtgaaatc-3'
2. *Taq* polymerase: AmpliTaq DNA polymerase (Perkin Elmer-Cetus, Branchburg, NJ).
3. PCR mix: The 50-μL reactions contain the following: 10 mM Tris-HCl, pH 8.0, 50 mM KCl, 1 mM MgCl$_2$, 200 μM of each dNTP (dATP, dGTP, dCTP, TTP), 165 nM of each primer, 1 U of *Taq* polymerase and 50 ng human leucocyte DNA.
4. Horizontal gel acrylamide stock: 29.1 g acrylamide (Bio-Rad, Hercules, CA), 0.9 g piperazine diacrylamide (Bio-Rad). Dissolve in 50 mL of water and make up to 100 mL with water. Store at 4°C in the dark. **Note:** Acrylamide is a neurotoxin.

5. Tris-SO$_4$ buffer: 80 g Trizma base, 62 mL 0.5M H$_2$SO$_4$, 200 mL H$_2$O. Adjust to pH 9.0 with Trizma base and make up to 400 mL.
6. Horizontal gel mix: 3.0 mL Horizontal gel acrylamide stock, 6.2 mL glycerol (20%), 8.2 mL Tris-SO$_4$ buffer, 10 mg ammonium persulfate (Boehringer Mannheim, Mannheim, Germany), 10 µL tetramethylethylenediamine (TEMED) (Boehringer).
7. Tris-borate buffer: 15.72 g Trizma base, 2.19 g boric acid. Dissolve with 150 mL water and adjust to pH 9.0 with Trizma base. Make up to 250 mL with water.
8. Agarose gel strips: 2% agarose in Tris-borate buffer. Boil the agarose in buffer (200 mL) and pour into a casting tray to form a gel 25 × 15 × 0.5 cm. Strips approx 1 cm wide are used for the electrode buffer.
9. Mylar film: Gel Bond PAG film (FMC Bioproducts, Rockland, ME).
10. Ethidium bromide solution: 10 µg/mL made in water. **Note:** Ethidium bromide is a mutagen.
11. Elution buffer: 10 mM Tris-HCl, pH 8.0, 100 mM KCl, 3 mM MgCl$_2$. Autoclave before use.

2.2. Sequencing Reactions

1. Sequencing master mix: 20 mM Tris-HCl, pH 8.0, 100 mM KCl, 3 mM MgCl$_2$, 330 nM each primer, 5 U of *Taq* polymerase/5 µL master mix.
2. Termination mixes: Each termination mix has 20 µM dGTP, 20 µM dATP, 20 µM TTP and 20 µM dCTP (Boehringer) together with the following: G mix: 60 µM ddGTP; A mix: 800 µM ddATP; T mix: 1200 µM ddTTP; C mix: 400 µM ddCTP (Boehringer).
3. Formamide stop solution: 10 mL formamide, 10 mg xylene cyanol, 10 mg bromophenol blue, 10 mM ethylene diamine tetra-acetic acid (EDTA), pH 8.0.
4. Repel Silane (Pharmacia, Uppsala, Sweden).
5. TBE: 81.2 g Tris base. 13.7 g boric acid, 5 g disodium EDTA in 500 mL water.
6. 40% acrylamide/bis-acrylamide solution (19:1, 5% C) (Bio-Rad).
7. 6% Acrylamide/urea stock: 230 g urea, 75 mL 40% acrylamide solution, 50 mL TBE. Dissolve the urea and make up to 500 mL with water.
8. Sequencing gel mix: 50 mL of 6% stock acrylamide/urea solution, 250 µL 10% ammonium persulfate and 50 µL TEMED.

2.3. DNA Transfer Procedures

DNA transfer papers: The following pieces of blotting paper (standard stationery type) and immobilon-S membrane (Millipore) are required for each gel transfer. Blotting paper #1: 16 × 33 cm. Immobilon-S membrane 18 × 34 cm soaked in 0.5X TBE. Blotting paper #3: 20 × 34 cm. Blotting paper #4: 20 × 34 cm soaked in 0.5X TBE.

2.4. Hybridization and Detection

1. Terminal transferase mix: Buffer (as supplied by Promega, Madison, WI), 20 µM dUTP-biotin, 60 µM TTP, 2 µM probe, 20 U terminal transferase (Promega).
2. 100 mM ethylene diamine tetraacetic acid (EDTA), pH 8.0.

3. Prehybrization solution: 7.3 g NaCl, 2.41 g Na_2HPO_4, 1.25 g $NaH_2PO_4 \cdot 2H_2O$, 49.89 g sodium lauryl sulfate (SDS), 100 g polyethylene glycol 6000. Make up to 1 L with water. Store at 25°C.

4. Blocking solution: 7.3 g NaCl, 2.41 g Na_2HPO_4, 1.25 g $NaH_2PO_4 \cdot 2H_2O$, 49.89 g SDS. Make up to 1 L with water. Store at 25°C.

5. Wash solution I: 1:10 dilution of blocking solution.

6. Wash solution II: 100 mM Tris-HCl, pH 9.5, 100 mM NaCl, 10 mM $MgCl_2$. Store at 4°C and dilute 1:10 for use.

7. Buffer S: 10 mM sodium phosphate, pH 7.2, 150 mM NaCl, 0.05% sodium azide.

8. Streptavidin solution: Add 1 mL of buffer S to 3 mg powdered streptavidin (as supplied in the Millipore Plex Luminescent Kit [Bedford, MA]). Store at 4°C.

9. Lumigen in diluent: (Millipore Plex Luminescent Kit). Make up fresh immediately before use.

10. X-ray film: HT-U Medical X-Ray film (Agfa Curix).

3. Methods

3.1. PCR Conditions

1. Mix DNA samples with PCR components; amplify DNA. PCR cycling conditions in a Perkin Elmer Cetus Thermal Cycler are 30–40 cycles of 94°C for 10 s, 55°C for 1 min. 72°C for 1 min (*see* Note 2).

2. *See* Note 3: steps 2 to 7 may not be necessary.

3. Pour a polyacrylamide gel. Pour the gel mix (5% polyacrylamide crosslinked with piperazine diacrylamide) onto a glass plate edged with plastic (0.4 mm thick). Layer a sheet of Gelbond plastic onto the liquid and allow the gel to set for 1 h. Place agarose strips containing Tris-borate buffer at either end of the gel.

4. Load the PCR samples (40 µL of PCR product) onto filter paper wicks (5 × 40 mm) and place onto the surface of the polyacrylamide gel.

5. Carry out electrophoresis for 2 h at between 1.8 and 6 W (at a constant 30 mAmps) across 12 cm.

6. Stain the gels in a solution of 10 µg ethidium bromide/mL water for 30 min and destain in water for 10 min. The bands are visualized with UV light. Cut out the bands from the gel with a sterile scalpel blade and then transfer to an Eppendorf tube.

7. Elute the DNA from the gel in 40 µL of elution buffer at 4°C for 16 h.

3.2. Sequencing Reactions

1. Add the purified PCR product (8 µL) to 5 µL sequencing master mix and 8 µL elution buffer. Add 5 µL of this mix to 5 µL of each of the termination mixes.

2. Take the mixture through 10 cycles of sequencing according to the PCR cycling conditions.

3. Stop the reaction by adding 4 µL of formamide stop solution (*see* Note 4 to make the sequencing time effective).

4. A Bio-Rad sequencing apparatus is used for the gel electrophoresis. Coat one glass plate with Repel Silane to ensure clean separation of the gel from the glass

plate. Pour the sequencing gel mix into the Bio-Rad sequencing apparatus as per the normal gel preparation procedure.

5. Prerun the gel at 52 W for 1 h (temperature 50–55°C) prior to loading the samples.
6. Heat the samples to 95°C for 2 min and keep at 72°C for 30 min prior to loading.
7. Load 6 μL of each sample onto the sequencing gel and carry out electrophoresis at 52 W for 2 h.

3.3. Transfer of the DNA to Nylon Membrane

1. Pry the glass plates apart and place a dry piece of blotting paper on top of the gel. Trim the gel to the size of the paper.
2. When the gel is peeled from the glass plate, it remains fixed to the blotting paper. Presoak Immobilon membrane (*see* Note 5) in 0.5X TBE and apply to the gel surface. Ensure that no bubbles are trapped beneath the membrane. Trim the edges of the membrane/gel/blotting paper sandwich to ensure the edges are even.
3. Place the membrane sandwich down onto a piece of dry blotting paper on a vacuum drying apparatus (Bio-Rad Model 543 Gel Drier) and place blotting paper wet with 0.5X TBE on top.
4. Transfer the DNA at room temperature by the application of reduced pressure (approx 10 Pa.) for 1 h (*see* Note 6).
5. Peel the membrane from the gel/blotting paper sandwich and dry at 60°C for 1 h.
6. Expose the membrane to UV light (placed on the surface of a 302 nm UV transilluminator) for 2 min to fix the DNA to the membrane (*see* Note 7). The membrane may be stored between blotting paper at this stage.
7. Oligonucleotide probes are labeled with dUTP-biotin using terminal transferase in 50 μL of the terminal transferase mix (*see* Note 8). The mixture is incubated at 37°C for 30 min and the reaction stopped with 10 μL 100 m*M* EDTA, pH 8.0.

3.4. Hybridization and Detection

The following hybrization, washing, and detection steps are performed in a rotating bottle apparatus to minimize background and volumes required (*see* Note 9). The procedure uses a Millipore Plex Luminescent Kit but laboratories could make up the individual components.

1. Add 60 mL prehybridization solution to a membrane in a plastic bottle and incubate with rotation at 50°C for 5 min. Discard the solution.
2. Add 60 μL labeled probe to 30 mL prehybridization solution and add to the bottle. Incubate with rotation at 50°C for 1 h. Discard the solution.
3. Wash the membrane in the bottle with 150 mL wash solution I for 3 min at room temperature. Repeat this step twice more. Discard the solution.
4. Add 60 mL blocking solution and incubate with rotation for 5 min at room temperature. Discard the solution.
5. Add 5.6 μL streptavidin solution to 17 mL blocking solution. Incubate with rotation for 5 min at room temperature. Discard the solution.
6. Add 150 mL wash solution I and incubate with rotation at room temperature for 5 min. Discard the solution and repeat this step.

7. Add 60 mL blocking solution and incubate with rotation at room temperature for 5 min. Discard the solution.
8. Add 2.2 µL biotinylated alkaline phosphatase (Millipore kit) in 17 mL blocking solution and incubate with rotation at room temperature for 5 min. Discard the solution.
9. Add 150 mL wash solution II and incubate with rotation at room temperature for 5 min. Discard the solution and repeat this step.
10. Add 100 µL lumigen in 10 mL diluent (Millipore kit) and incubate with rotation at room temperature for 5 min.
11. Place the membrane between two plastic sheets and expose to X-ray film for 45–90 min (time depends on the efficiency of the probing). Develop the film as normal.
12. Transfer the membrane to a plastic bottle. Add 150 mL wash solution, preheated to 65°C, and incubate with rotation at 65°C for 15 min. Discard the solution
13. Add 150 mL wash solution I and incubate with rotation at room temperature for 10 min. Discard the solution and repeat this step. The membrane may be reprobed or stored in wash solution I at room temperature.

4. Notes

1. The probes used for mitochondrial DNA analysis are from sequences found within the PCR product. The use of internal probes may help to reduce the number of spurious bands appearing on the gels. The published procedure *(1)* did not require internal probes and it details the use of probes that were complementary to the primers.
2. The procedure is best suited for fragments of the order of 400 bases or less. The PCR of longer fragments appears to be less efficient although the newer PCR systems (Cetus) designed specifically for the synthesis of long fragments may improve efficiency.
3. PCR should be attempted at the lowest possible dNTP concentration. In the majority of cases, 20 µ*M* will produce a product suitable for sequencing without the need for gel purification. In some instances, however, spurious bands cannot be removed by optimizing the PCR, or the PCR is inefficient and we recommend purification of the product on a gel after PCR at higher dNTP concentrations. This has the dual effect of removing excess dNTPS and any spurious products.
4. A number of PCR products can be placed into the sequencing reactions. The only limitation is the volume that can be loaded onto the sequencing gel. By sequencing a number of DNA fragments in the one reaction, the electrophoresis time per PCR product is dramatically reduced and, hence, improves the efficiency of the overall sequencing procedure.
5. The best membrane appears to be Immobilon-S membrane from Millipore. Other membranes appear to produce higher background problems and would require changes to the blocking protocols developed by Millipore.
6. DNA can be transferred to a membrane by the classical Southern blotting technique, but in our hands this is inefficient. Other transfer procedures (e.g., electrophoretic) have not been attempted.

7. Fixation of the DNA to the membrane is critical. As each UV light source is likely to vary, the exposure time should be empirically determined.
8. The protocol allows for two probings per day. This could be increased by the use of probes labeled with alkaline phosphatase as the number of blocking/washing steps would be reduced.
9. Standard hybridization procedures in plastic bags produced high background. Hybridization in a rotating bottle apparatus uses less materials and clearer backgrounds. The membranes should be handled at the edges, as any imprint will appear on the final X-ray film.

References

1. Douglas, A. M., Georgalis, A. M., and Atchison, B. A. (1993) Direct sequencing of double-stranded PCR products incorporating a chemiluminescent detection procedure. *BioTechniques* **14,** 824–828.
2. Gyllensten, U. B. and Erlich, H. A. (1988) Generation of single-stranded DNA by the polymerase chain reaction and its application to direct sequencing of the HLA-DQA locus. *Proc. Natl. Acad. Sci. USA* **85,** 7652–7656.
3. Kusukawa, N., Uemori, T., Asad, K., and Kato, L. (1990) Rapid and reliable protocol for direct sequencing of material amplified by the polymerase chain reaction. *BioTechniques* **9,** 66–72.
4. Engelke, D. R., Hoener, P. A., and Collins, F. S. (1988) Direct sequencing of enzymatically amplified human genomic DNA. *Proc. Natl. Acad. Sci. USA* **85,** 544–548.
5. Ruano, G. and Kidd, K. K. (1991) Coupled amplification and sequencing of genomic DNA. *Proc. Natl. Acad. Sci. USA* **88,** 2815–2819.

8

Direct DNA Sequencing
of PCR Products Using Magnetic Beads

Joakim Lundeberg, Bertil Pettersson, and Mathias Uhlén

1. Introduction

The use of magnetic particles in many fields of biochemistry, molecular biology, and medicine has been well documented, and several magnetic particles are now available for diagnostic and cell-separation purposes. The solid-phase approaches have improved robustness by their increased reproducibility with accompanying higher yields. Furthermore, automation has been facilitated since reaction buffers and additional reagents can be rapidly changed without centrifugation or precipitation steps. The introduction of magnetic particles has proven to be an attractive alternative to traditional means of performing DNA sequencing.

Template preparation for DNA sequencing has relied on cloning of the target sequence into phage or plasmid vectors. After cultivation, sufficient amounts of target DNA can be prepared to enable chain-termination DNA sequencing as described by Sanger and coworkers (1). This indirect method, using cloning to prepare DNA for sequencing, has become more efficient by the use of polymerase chain reaction (PCR). By the design of PCR primers containing restriction handles, a rapid and efficient cloning procedure into suitable sequencing vectors (2) is achieved. However, thermostable *Taq* DNA polymerase used in PCR lacks proofreading activity, and therefore, multiple clones may have to be sequenced and analyzed to find the consensus sequence.

Alternatively, an in vitro amplification can be performed on genomic DNA that allows for direct DNA sequencing of the PCR products. Thus, both time-consuming cloning steps as well as control sequencing can be avoided. Therefore, the DNA sequence of a sample can be determined rapidly and will represent the sequence of the sample prior to amplification, since the errors

From: *Methods in Molecular Biology, Vol. 65: PCR Sequencing Protocols*
Edited by: R. Rapley Humana Press Inc., Totowa, NJ

produced by *Taq* polymerase will not significantly contribute to the resulting signal. A clear disadvantage is the need to remove excess PCR primers, enzyme, and nucleotides. This can usually be solved by a precipitation step, spin columns, or HPLC. A more serious disadvantage is the competition between primer annealing and reannealing of the complementary strand, which causes severe DNA sequencing interpretation problems.

Monodisperse paramagnetic beads coated with streptavidin can be used to capture and purify biotinylated PCR products. Thereby many of the problems in the preparation of sequencing templates are circumvented and a robust system for DNA sequencing (Fig. 1) *(3,4)* is achieved. The method allows for complete removal of one strand, nucleotides, DNA polymerase, and PCR primers in a few simple steps. This procedure results in a pure and single-stranded template, and in addition, reassociation problems of the complementary strand are minimized. The principle of the solid-phase approach is the use of the strong interaction between biotin and streptavidin, which is extremely strong ($K_d = 10^{-15}M$), temperature-stable (up to 80°C), and withstands alkali treatment (0.1M NaOH). The most preferred manner to introduce a biotin label into a double-stranded DNA fragment is to have one of the PCR primers biotinylated. Biotinylation of primers has been significantly simplified by the introduction of biotin phosphoramidites, which enables a direct coupling onto the 5'-end during primer synthesis (*see* Note 1). The capture and immobilization are accomplished by incubation of the biotinylated PCR product with the streptavidin-coated magnetic beads for a few minutes. The captured DNA is denatured into two single strands by the addition of 0.1M NaOH. This results in the elution of the nonbiotinylated strand into the supernatant, whereas the biotinylated strand remains immobilized to the bead surface. The single-stranded template in the supernatant can be recovered by the use of a magnet and subsequent removal of the eluate into a separate tube for neutralization. The benefit of this method is that all reaction components are removed, including the complementary strand, enabling optimal sequencing conditions with no reannealing problems. Note that after neutralization, the eluted nonbiotinylated strand can also be used as a template. Furthermore, the solid-phase approach enables the development of integrated and automated methods for routine sequencing, facilitated by the defined and predictable behavior of the monodisperse magnetic beads *(3,4)*.

All of the available DNA polymerases are suitable for solid-phase DNA sequencing, such as, T7, Klenow, *Taq*, *Tth*, *Bst*, and so forth. However, the thermostable enzymes poorly incorporate dideoxynucleotides in an inconsistent pattern in comparison with native deoxynucleotides *(5,6)*. The variation in relative signal intensities, caused by *Taq* polymerase, makes the interpretation by the software algorithm used in automated fluorescent electrophoresis units

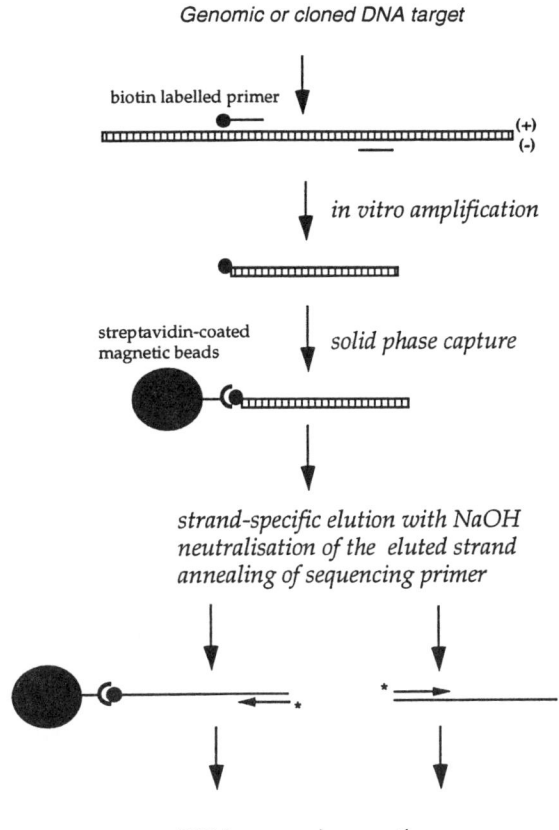

Fig. 1. The solid-phase sequencing concept.

more difficult *(7)*. Despite the disadvantage of nonuniform signal intensities, the convenience of rapid sequence generation by temperature cycling ("cycle sequencing") has made *Taq* DNA polymerase useful for sequencing of PCR products and cloned target DNA *(8)*. In clinical applications however, the T7 DNA polymerase (Sequenase™ Version 2.0) is the enzyme of choice because of a high processivity, which results in more uniform bands compared to the other polymerases. Peak uniformity is important for accurate base-calling, especially for the detection of heterozygosity in genomic material *(6,9)*.

The automation of DNA sequencing in recent years has opened new application fields, especially within clinical medicine. The most important developments toward automated routine use are the introduction of PCR, and the replacement of isotopic labeling by fluorescent dyes and on-line monitoring of the DNA sequence. The fluorescent sequencing bands are excited by a laser beam and detected in the gel during electrophoresis. There are three main methods to obtain labeled dideoxy DNA fragments:

1. Fluorescent labeled primer.
2. Fluorescent labeled dideoxy chain terminators.
3. Fluorescent labeled dATPs.

These alternatives enable many flexible nonradioactive sequencing solutions. The commercially available instruments also enable reliable quantitation of polymorphic and heterozygous positions with the different software packages *(10–12)*. Note, the manual protocols described below have also been used as the basis for developing semiautomated sequencing systems. The template preparation and the sequencing reactions have been implemented on a Beckman Biomek 1000 work station *(4)* and on an ABI Catalyst work station *(13)*.

The following sections describe protocols for:

- Amplification of plasmid inserts and genomic DNA targets: General vector primers can be used to amplify inserts in plasmids, such as pUC, pBluescript, pEMBL, pGEM, and so forth. A simplified protocol for template preparation of bacterial colonies has therefore been included in combination with a protocol for amplification with universal primer sets containing one biotinylated primer and one nonbiotinylated primer. Amplification of a target gene in genomic DNA can also be performed using a designed primer set with one biotinylated primer (*see* Note 1). However, obviously, the choice of primers influences the conditions used in the amplification reaction and must be adapted to fit the primer pair chosen for the amplification.
- Capture of the PCR product and strand separation: The biotinylated PCR products are directly immobilized to prepared streptavidin-coated paramagnetic beads. Denaturation of the immobilized double-stranded DNA on the beads, followed by elution of the nonbiotinylated strand, yields two single-stranded DNA templates: one immobilized on the beads and the other in the supernatant.
- Solid-phase sequencing protocols: Two alternative sequencing protocols are described below. The first is suitable for manual sequencing using radioactive labeled primers or for automated DNA sequencing using the Pharmacia ALF system. The second system employs four dye fluorescent labeled primers and requires access to an automated DNA sequencer from Applied Biosystems Inc. The sequencing primers to be used can either be custom-designed (complementary to a sequence inside the target DNA being amplified) *(3)*, a universal primer complementary to a sequence introduced by a "handle" sequence in one of the PCR primers *(14,15)*, or one of the primers used in the PCR *(10,16)*. The sequencing primers used for plasmid vectors depends on the choice of PCR primers. PCR primer set A: the immobilized strand can be sequenced with the universal sequencing primer. Alternatively, the immobilized strand can be sequenced with the T3 or the SP6 promoter primers. The eluted (supernatant) strand can be sequenced using the reverse sequencing primer or the T7 promoter primer. PCR

primer set B: the immobilized strand is sequenced with the reverse sequencing primer or the T7 promoter primer. The supernatant strand can be sequenced using the universal sequencing primer, T3, or the SP6 promoter primers.

2. Materials

1. Colony lysis buffer: 100 mM Tris-HCl, pH 8.3 (20°C), 20 mM MgCl$_2$, 500 mM KCl, 1.0% Tween 20.
2. 10X PCR buffer: 100 mM Tris-HCl, pH 8.3 (20°C), 20 mM MgCl$_2$, 500 mM KCl, 1% Tween 20.
3. Thermocycler.
4. AmpliTaq polymerase (Perkin-Elmer Cetus, Norwalk).
5. Nucleotide solution (2 mM of each dNTP).
6. General PCR vector primers solution containing two primers each having a concentration of 2.5 μM. Primer set A: 5'-biotin-GCTTCCGGCTCGTATGTT GTGTG-3'; 3'-GCGGAACGTCGTGTAGGGGGAAA-5'; primer set B: 5'-GCTT CCGGCTCGTATGTTGTGTG-3'; 3'-GCGGAACGTCGTGTAGGGGGAAA-biotin-5'.
7. Dynabeads M-280 streptavidin (10 mg/mL) (Dynal AS, Oslo, Norway).
8. Neodymium-iron-boron magnet (MPC Dynal AS, Oslo, Norway).
9. Washing/binding solution (10 mM Tris-HCl, pH 7.5 [20°C], 1 mM EDTA, 2M NaCl).
10. 1X TE buffer (10 mM Tris-HCl, pH 7.5 [20°C], 1 mM EDTA)
11. 0.10M NaOH, freshly made (*see* Note 2)
12. 0.333M HCl, freshly made (*see* Note 3)
13. T buffer (0.75M Tris-HCl, pH 7.5 [20°C]).
14. Water baths or incubators at 0, 37, and 65°C.
15. A polyacrylamide sequencing gel and electrophoresis equipment or automated sequencers (Pharmacia ALF, Uppsala, Sweden or ABI Automated Sequencer, Foster City, CA).
16. Annealing buffer for single-labeled sequencing primer: 280 mM Tris-HCl, pH 7.5 (20°C), 100 mM MgCl$_2$. The label can be either ^{32}P for radioactive sequencing or one dye fluorescent label, such as fluorescein isothiocyanate (FITC) (Pharmacia, Uppsala, Sweden).
17. Annealing buffer for four dye-labeled sequencing primers: 56 mM Tris-HCl, pH 7.5 (20°C), 0.05% Tween 20, 20 mM MgCl$_2$. The general dye-labeled sequencing primers are obtained from ABI Inc., Foster City, CA.
18. Extension buffer: 300 mM citric acid, pH 7.0 (20°C), 318 mM DTT, 40 mM MnCl$_2$.
19. T7 DNA polymerase with enzyme dilution buffer (Pharmacia, Uppsala, Sweden).
20. Four nucleotide mixes each containing 40 mM Tris-HCl, pH 7.5 (20°C), 50 mM NaCl, 1.0 mM of each dNTP and 5.0 μM of one specific ddNTP. Thus, the "A" mix contains 5 μM ddATP, the "C" mix contains 5 μM ddCTP, the "G" mix contains 5 μM ddGTP, and the "T" mix contains 5 μM ddTTP.
21. Stop solution: Shake 100 mL formamide with 5 g Amberlite MB-1 resin and 300 mg dextran blue for 30 min. Filter through 0.45-μm pore-size filter.

3. Methods

3.1. Colony Template Preparation

Dispense 10 μL of lysis buffer into each tube. Take a part of one colony from the cultivation plate using a toothpick and place it in the PCR tube. Incubate in the thermocycler at 99°C for 5 min and immediately place the tube on ice.

3.2. Amplification of Plasmid Inserts and Genomic DNA Targets

1. Prepare a PCR-master mix in a microcentrifuge tube: 5 μL 10X PCR buffer, 2 μL (5 pmol) primers (*see* Notes 1 and 4), 5 μL dNTP solution, 1 U AmpliTaq, and sterile water to 50 μL.
2. Use 1–2 pmol of template DNA (or 5 μL of the colony lysate).
3. Cover with 25 μL of light mineral oil, if necessary. For amplification of plasmid inserts, cycle as follows: 96°C for 30 s; 72°C for 0.5–3.0 min (depending on the length of target sequence) for 30 cycles. A final extension step at 72°C for 10 min is strongly suggested.
4. Analyze the PCR product (2–3 μL) on an agarose gel.

3.3. Preparation of Streptavidin-Coated Magnetic Beads

1. Resuspend the beads by pipeting. Use 20 μL (200 μg) of resuspended beads/PCR template, and pipet the suspended beads into a clean 1.5-mL microcentrifuge tube. The beads may be washed in bulk for the total number of PCR templates that need to be purified.
2. Place the tube in the magnetic holder, and allow for beads to adhere to the magnet at the side of the wall. Remove the supernatant using a pipet (do not remove the tube from magnetic holder).
3. Add an equal volume of washing/binding solution, and gently pipet to suspend.
4. Repeat step 2 and 3. Allow the beads to adhere to the side of the tube and remove the supernatant.
5. Resuspend the beads in washing/binding solution using twice the original volume (i.e., 40 μL). The bead concentration is now 5 μg/μL.

3.4. Capture of the PCR Product and Strand Separation

1. Take 45 μL of the total 50 μL PCR amplification reaction to a fresh 1.5-mL microcentrifuge tube, and add 40 μL of the prewashed Dynabeads (*see* Note 4).
2. Incubate at room temperature for 15 min. Mix during the immobilization reaction once or twice by gentle pipeting or tapping.
3. Collect the beads by moving the vials to the magnetic holder, and remove the supernatant with a pipet.
4. Wash the beads once with 50 μL washing/binding solution (*see* Note 5).
5. Wash once with 50 μL 1X TE buffer. Remove the 1X TE buffer carefully. Avoid droplets on the walls and the bottom of the tube.
6. Resuspend the beads in **exactly** 10 μL 0.10*M* NaOH (*see* Note 2).
7. Incubate at room temperature for 5 min.

8. Collect the beads (now with only ssDNA attached) by placing the tube in the magnetic holder, and transfer the 10 μL of NaOH supernatant (containing the nonbiotinylated strand) to a clean tube. Neutralize the NaOH supernatant with 3 μL 0.333M HCl (*see* Note 3) and mix **immediately.** Add 2 μL of T buffer, and save the supernatant strand.
9. Wash the beads once with 50 μL 0.1M NaOH, once with 50 μL washing/binding solution, and once with 50 μL 1X TE buffer. Remove the 1X TE carefully, without leaving any droplets.
10. Resuspend the beads in 15 μL of sterile water (or the appropriate buffer for the sequencing protocol to follow).

3.5. Solid-Phase DNA Sequencing Using T7 DNA Polymerase and One Dye-Labeled Primer

1. Add template DNA (beads with immobilized single-strand DNA or eluted single-strand DNA) to a fresh tube (*see* Note 6). Adjust the volume (with sterile water) to 15 μL.
2. Add 2 μL (1 pmol) labeled primer.
3. Add 2 μL of annealing buffer, and mix gently with a pipet. Incubate at 65°C for 10 min. Mix gently and leave to cool at room temperature for at least 10 min. Mix two times during cooling.
4. Add 1 μL of extension buffer, and mix gently.
5. Dilute the T7 DNA polymerase to 1.5 U/μL using **cold** dilution buffer; 2 μL of this diluted stock solution will be required for each template, and keep the tube with diluted stock solution (1.5 U/μL) on ice.
6. Label four new tubes "A," "C," "G," and "T" (*see* Note 7). Dispense 2.5 μL of the corresponding dNTP/ddNTP sequencing mixes into the tubes.
7. Warm the dispensed nucleotide sequencing mixes at 37°C for at least 1 min.
8. Add 2 μL of the T7 polymerase diluted stock solution (from step 5) to the template mixture (from step 4), and mix gently. Immediately add 4.5 μL of this mixture to each of the preincubated nucleotide sequencing mixes.
9. Incubate at 37°C for 5 min.
10. Add 5 μL of stop solution to each reaction, and mix gently.
11. Incubate at 95°C for 5 min, and then put the tubes on ice.
12. Load the samples onto the polyacrylamide sequencing gel.

3.6. Solid-Phase DNA Sequencing Using T7 DNA Polymerase and Four Dye Fluorescent Primers for Plasmid Sequencing

1. Dissolve the immobilized single-stranded template in 19 μL of annealing buffer in a microcentrifuge tube (*see* Note 6).
2. Dilute the T7 DNA polymerase to 0.7 U/μL using **cold** dilution buffer; 6 μL of this diluted "stock solution" will be required for each template, and keep the tube with diluted stock solution (0.7 U/μL) on ice.
3. Aliquot the paramagnetic beads with the immobilized template DNA into four tubes labeled "A," "C," "G," and "T." Add ABI's plasmid fluorescent sequencing

primer as follows: for "A" and "C," take 3 µL beads (in annealing buffer), 1 µL primer, and 0.5 µL extension buffer; for "G" and "T," take 6 µL beads (in annealing buffer), 2 µL primer and 1 µL extension buffer. Mix gently, and heat to 65°C for 10 min. Mix gently, and leave to cool at room temperature for at least 10 min, mixing two or three times during cooling.

4. Add an appropriate volume of the relevant dNTP/ddNTP sequencing mixes ("A," "C," "G," or "T") to each tube as indicated: for "A" and "C" take 1.5 µL nucleotide mixtures; for "G" and "T," take 3 µL nucleotide mixtures (*see* Note 7). Keep the samples on ice while doing this.

5. Add the T7 DNA polymerase diluted stock solution, as indicated: for "A" and "C," take 1 µL T7 DNA polymerase diluted stock solution; for "G" and "T," take 2 µL T7 DNA polymerase diluted stock solution. Incubate at 37°C for 5 min.

6. Stop the reaction by adding 40 µL ice-cold 10X TE to each tube, and place the tubes on ice.

7. Pool the "A," "C," "G," or "T" reactions for each sample, and collect the beads using the magnet. Discard the supernatant. Wash the beads once with 1X TE buffer, and discard the supernatant. Resuspend the beads in 5 µL loading buffer (*see* Note 8).

8. Prior to loading the samples onto the ABI automated DNA sequencer, heat the samples to 95°C for 2 min and place on ice.

4. Notes

1. Specific primers for the genomic target gene must be designed and synthesized. A nested primer procedure is strongly recommended to yield a clean product (14–16). Biotinylated oligonucleotides can be obtained from several commercial sources offering oligonucleotide synthesis services. It is of great importance that the biotinylated oligonucleotide be purified from unbound biotin, preferably by reverse-phase FPLC or HPLC, since free biotin will occupy binding sites on the beads and reduce the binding capacity of biotinylated PCR products. DNA synthesis can use available biotin phosphoramidites for 5'-end biotinylation of oligonucleotides (Biotin-ON™ Phosphoramidites, Clontech Laboratories Inc., Palo Alto, CA; Biodite, Pharmacia, Uppsala, Sweden). Aliquots of labeled primers should be stored at –20°C (avoid repeated freeze-thawing).

2. Important: Use a $1.000 \pm 0.005M$ volumetric solution of NaOH, and dilute this to $0.1M$. Aliquot this and store at –20°C.

3. Important: use a $1.000 \pm 0.005M$ volumetric solution of HCl, and dilute this to $0.1M$. Aliquot this and store at –20°C.

4. The binding capacity of the beads is fragment-length-dependent. Reduced binding capacity for longer DNA fragments is probably caused by steric hindrance on the bead surface. However, a prolonged binding time increases the degree of capture. For example, a long product can be incubated for 60 min at 43°C or, alternatively, overnight at room temperature. In addition, one may double the amount of beads to decease the effect of steric hindrance on the bead surface. Another option is to change the binding buffer to contain $3M$ LiCl instead of $1M$

NaCl, which gives a slight improvement in binding capacity. In addition, a threefold excess of nonbiotinylated primer may be used to drive the complete extension of the biotinylated fragment to minimize saturation of free biotin primers onto the streptavidin surface.

5. The immobilized DNA can be stored at 4°C for several weeks.
6. If many samples are to be analyzed, a microtiter plate might be more convenient.
7. 7-Deaza-dGTP (c7dGTP) is preferred instead of dGTP for resolving band compression during electrophoresis.
8. The sequencing reactions may be stored at –20°C if not loaded immediately. Prior to loading, heat the samples to 95°C for 2 min, and place on ice.

Acknowledgments

This work was supported by the Göran Gustafsson Foundation for Research in Natural Sciences and Medicine. The authors thank Johan Wahlberg and Thomas Hultman for considerable help with the protocols.

References

1. Sanger, F., Nicklen, S., and Coulson, A. R. (1977) DNA sequencing with chain-terminating inhibitors. *Proc. Natl. Acad. Sci. USA* **74(12),** 5463–5467.
2. Scharf, S. J., Horn, G. T., and Erlich, H. A. (1986) Direct cloning and sequence analysis of enzymatically amplified genomic sequences. *Science* **233,** 1076–1078.
3. Hultman, T., Ståhl, S., Hornes, E., and Uhlén, M. (1989) Direct solid phase sequencing of genomic and plasmid DNA using magnetic beads as solid support. *Nucleic Acids Res.* **17,** 4937–4946.
4. Hultman, T., Bergh, S., Moks, T., and Uhlén, M. (1991) Bidirectional solid-phase sequencing of in vitro-amplified plasmid DNA. *BioTechniques* **10(1),** 84–93.
5. Innis, M. A., Myambo, K. B., Gelfand, D. H., and Brow, M. A. D. (1988) DNA sequencing with *Thermus aquaticus* DNA polymerase and direct sequencing of PCR-amplified DNA. *Proc. Natl. Acad. Sci. USA* **85,** 9436–9440.
6. Leren, T. P., Rødningen, O. K., Røsby, O., Solberg, K., and Berg, K. (1993) Screening for point mutations by semi-automated DNA sequencing using Sequenase and magnetic beads. *BioTechniques* **14(4),** 618–623.
7. Khurshid, F. and Beck, S. (1993) Error analysis in manual and automated DNA sequencing. *Anal. Biochem.* **208,** 138–143.
8. Carothers, A. M., Urlaub, G., Mucha, J., Grunberger, D., and Chasin, L. A. (1989) Point mutation analysis in mammalian gene: Rapid preparation of total RNA, PCR amplification of cDNA, and *Taq* sequencing by a novel method. *BioTechniques* **7,** 494–499.
9. Hedrum, A., Pontén, F., Ren, Z., Lundeberg, J., Pontén, J., and Uhlén, M. (1994) Sequence-based analysis of the human p53 gene based on microdissection of tumor biopsy samples. *BioTechniques* **17,** 1–9.
10. Wahlberg, J., Albert, J., Lundeberg, J., Cox, S., Wahren, B., and Uhlén, M. (1992) Dynamic changes in HIV-1 quasispecies from azidothymidine (AZT) treated patients. *FASEB J.* **6,** 2843–2847.

11. Leitner, T., Halapi, E., Scarletti, G., Rossi, P., Albert, J., Fenyö, E.-M., and Uhlén, M. (1993) Analysis of heterogeneous viral populations by direct DNA sequencing. *BioTechniques* **15(1),** 120–127.
12. Larder, B. A., Kohli, A., Kellam, P., Kemp, S. D., Kronick, M., and Henfrey, R. D. (1993) Quantitative detection of HIV-1 drug resistance mutations by automated DNA sequencing. *Nature* **365,** 671–673.
13. Holmberg, A., Fry, G., and Uhlén, M. (1993) Automated preparation of DNA templates for sequencing on the ABI Catalyst robotic workstation, in *Automated DNA Sequencing and Analysis Techniques* (Venter, C., ed.), Academic, London, pp. 139–145.
14. Wahlberg, J., Albert, J., Lundeberg, J., Fenyö, E.-M., and Uhlén, M. (1991) Analysis of the V3 loop in neutralization-resistant human immunodeficiency virus type 2 variants by direct solid phase DNA sequencing. *AIDS Res. Hum. Retrov.* **7(12),** 983–990.
15. Wahlberg, J., Lundeberg, J., Hultman, T., and Uhlén, M. (1990) General colorimetric method for DNA diagnostics allowing direct solid phase genomic sequencing of the positive samples. *Proc. Natl. Acad. Sci. USA* **87,** 6569–6573.
16. Pettersson, B., Johansson, K.-E., and Uhlén, M. (1994) Sequence analysis of 16S rRNA from Mycoplasmas by direct solid-phase DNA sequencing. *Appl. Env. Microbiol.* **60(7),** 2456–2461.

9

Affinity Capture and Solid-Phase Sequencing of Biotinylated PCR Products

Anu Suomalainen and Ann-Christine Syvänen

1. Introduction

The time and labor required for generating nucleotide sequence information has been significantly reduced since the development of the polymerase chain reaction (PCR). PCR allows the production of sufficient amounts of DNA template in vitro, and consequently, the time-consuming cloning procedure to obtain large enough quantities of DNA for sequencing is avoided. The PCR has transformed nucleotide sequencing into a method that can be considered even in routine diagnostics for sequencing large numbers of individual samples.

There are, however, two major problems associated with the sequencing of PCR products. The first one originates from the fact that the PCR and the Sanger dideoxy-nucleotide sequencing methods are based on the same principle, i.e., a DNA-polymerase-catalyzed primer extension reaction with nucleotide triphosphates. The amount of the two primers in a typical PCR reaction is 50- to 100-fold compared to the standard amount of the single primer in a sequencing reaction, and the amount of deoxy-nucleotide triphosphates (dNTPs) in the PCR can be as much as 5000- to 10,000-fold compared to the amount of dNTPs in the labeling reaction or to the amount of the dideoxy-nucleotide triphosphates (ddNTPs) in the termination step of the sequencing procedure. Complete removal of the primers and the dNTPs from the PCR product is cumbersome, but essential for successfully performing the sequencing reactions. Another problem arises from the strong tendency of the two strands of a double-stranded PCR product to reanneal, which will compete with the annealing of the sequencing primer. Various approaches to cir-

From: *Methods in Molecular Biology, Vol. 65: PCR Sequencing Protocols*
Edited by: R. Rapley Humana Press Inc., Totowa, NJ

cumvent these problems have been taken *(1–3)*, and many of them are described in the present book.

We have solved the problems of direct sequencing of PCR products by employing affinity capture for purifying the sequencing template *(4)*. Our method is based on PCR with one biotinylated and one unbiotinylated primer, after which the PCR product, which will carry biotin in one of its strands, is captured on avidin-coated polystyrene beads, taking advantage of the strong interaction between biotin and avidin. Immobilization of the PCR product allows efficient and technically simple removal of the excess primers and dNTPs after PCR by washing, as well as the removal of the unbiotinylated strand by alkaline treatment. The immobilized single-stranded DNA template can then be sequenced by a standard dideoxy-nucleotide sequencing procedure.

The protocol presented below, which is routinely used in our laboratory, has been simplified from the method for sequencing affinity-captured PCR products originally developed for diagnostic purposes *(5)*. A similar kind of solid-phase sequencing method, in which streptavidin-coated magnetic beads are used as the solid support, was developed by Hultman et al. *(6)*. This method has been automated and has become widely used, especially combined with the use of automated DNA sequencers with fluorescent detection *(7)*. A recently devised comb-shaped streptavidin-coated manifold support offers the promise of further simplifying the technical performance of solid-phase sequencing reactions *(8)*.

The major advantage of the avidin-coated polystyrene beads in solid-phase sequencing is their small size (0.8–1.0 μM) and consequently their high biotin-binding capacity. The small size of these beads results in reaction kinetics approaching those of reactions in solution and in an extremely large binding surface (70 cm^2/mg of beads). The biotin-binding capacity of the polystyrene beads exceeds 2 nmol of biotinylated oligonucleotide/mg of beads, when the corresponding capacities for the streptavidin-coated magnetic beads or standard microtiter plates are 300 pmol/mg beads and 2–5 pmol/well, respectively *(9)*. An additional advantage of the polystyrene beads is their low price compared to that of magnetic ones. The drawback of these beads is the requirement of separation by centrifugation after the washing and denaturation steps.

2. Materials

All the reagents should be of standard molecular biology purity grade. Use sterile distilled or deionized water.

1. Facilities and reagents for PCR.
2. One of the PCR primers is biotinylated at its 5'-end during the oligonucleotide synthesis (biotin-phosphoramidite reagent, for example, Amersham, Buckinghamshire, UK, or Perkin Elmer/ABI, Foster City, CA), and the other one is not biotinylated,

which results in a PCR product with one biotinylated strand (*see* Note 1). In addition, an unbiotinylated sequencing primer is needed (*see* Note 2).

3. Avidin-coated polystyrene beads (Fluoricon®, avidin-polystyrene assay particles, 5% w/v, 0.99 μm, IDEXX Corp., Portland, ME). Store at 4°C (*see* Note 3).
4. PBS/Tween buffer: 20 mM sodium phosphate buffer, pH 7.5, 0.1% Tween 20.
5. TENT buffer: 40 mM Tris-HCl, pH 8.8, 0.1% Tween 20, 1 mM EDTA, 50 mM NaCl. Store at 4°C.
6. NaOH/Tween mixture: 50 mM NaOH, 0.1% Tween 20 (make fresh every 4 wk). Store at room temperature (about 20°C) (*see* Note 4).
7. Sequencing reagents: Any standard protocol for sequencing of single-stranded DNA templates, either with a [^{35}S] dNTP as label or with a [^{32}P] or fluorescently labeled primer, can be used. We use the following reagents and stock solutions for sequencing with [^{35}S] dATP as label (store all the reagents at –20°C):
   a. 5X T7 DNA polymerase buffer: 0.2M Tris-HCl, pH 7.5, 0.1M MgCl$_2$, and 25 mM NaCl.
   b. 0.1M dithiotreitol (DTT).
   c. Labeling mixture: 7.5 μM dGTP, dTTP, dCTP.
   d. [^{35}S] dATP (Amersham).
   e. T7 DNA polymerase (Pharmacia Biotech, Inc., Piscataway, NJ).
   f. Enzyme dilution buffer: 20 mM Tris-HCl, pH 7.5, 5 mM DTT, 100 μg/mL bovine serum albumin, and 5% glycerol.
   g. Four ddNTP termination mixtures: 80 μM dATP, dTTP, dCTP, dGTP, containing 8 μM of one ddNTP, in 50 mM NaCl.
   h. Formamide dye: 95% formamide (v/v), 20 mM EDTA, 0.05% bromphenol blue (w/v), 0.05% xylene cyanol (w/v).
8. Facilities for polyacrylamide gel electrophoresis.

3. Method
3.1. PCR for the Solid-Phase Sequencing

The PCR is undertaken according to standard protocols, with one biotinylated primer and the other one not biotinylated. We use 100-μL reaction volumes to obtain a sufficient amount of the PCR product. We have successfully sequenced products shorter than 2 kb (*see* Note 5). The annealing temperature and the amount of the template added should be optimized to result in a specific amplification product, i.e., one clearly visible band when 1/10 of the PCR product is analyzed by agarose gel electrophoresis stained with ethidium bromide. If the PCR is specific and efficient, no further purification of the PCR product is needed (*see* Note 6).

3.2. Affinity Capture
and Direct Sequencing of the PCR Product

1. Mix the following components: 10 μL of washed avidin-coated polystyrene beads, 25–60 μL of the PCR product (*see* Note 7), and TENT buffer to a total volume of 100 μL.

2. Capture the biotinylated PCR product on the avidin-coated beads for 30 min at 37°C.
3. Centrifuge the beads carrying the captured PCR product at 13,000g for 2 min, and remove the supernatant.
4. Denature the captured PCR product by adding 100 µL of the NaOH/Tween mixture, mix thoroughly, and incubate at room temperature for 5 min. Centrifuge as in step 3, and remove the supernatant. At this stage, the biotinylated strand of the PCR product remains bound to the beads.
5. Wash the beads twice by adding 200 µL of TENT buffer, mix well, centrifuge as in step 3, and carefully remove the supernatant (*see* Note 8).

The template is now ready for the sequencing reactions. Any protocol for sequencing single-stranded DNA templates can be applied (*see* Note 9). We use the following protocol:

6. Suspend the bead pellet carrying the captured template DNA with 7.5 µL of dH$_2$O, and add 0.5 µL (10 pmol) of the sequencing primer and 2 µL of 5X T7 DNA polymerase buffer. Mix and let the primer anneal at 37–42°C for 30 min (*see* Note 10).
7. It is convenient to prepare the dilutions for the following steps during the annealing step 6. Dilute the labeling mixture 1:4 with dH$_2$O. Dilute the T7 DNA polymerase with the enzyme dilution buffer to a concentration of 1.25 U/µL. Keep the dilutions on ice until used. Label Eppendorf tubes for each termination reaction ("C," "A," "T," "G"), and pipet 2.5 µL of the relevant ddNTP termination mixture to each tube.
8. Mix 1 µL of 0.1M DTT, 2 µL of diluted labeling mixture, and 0.5 µL of [^{35}S] dATP, and add this mixture to the tube containing the template and the annealed primer. Add 2 µL of diluted T7 DNA polymerase, and incubate the mixture for 4 min at room temperature (*see* Note 11).
9. Preheat the termination mixtures to 37°C for 1 min.
10. Transfer 3.5 µL of the sequencing reaction mixture to each preheated termination mixture, and incubate at 37°C for 3–4 min.
11. Stop the sequencing reactions by adding 4 µL of the formamide dye into each sample at room temperature, and store the samples at –20°C until use. The samples can be stored for at least 1–2 wk.
12. Denature the samples at 80°C for 3 min, and analyze 4 µL of each sample by denaturing 5–6% polyacrylamide gel electrophoresis (*see* Note 12) (*see* Chapter 1).

4. Notes

1. The efficiency of the 5'-biotinylation of an oligonucleotide on a DNA synthesizer is most often 80–90%. The biotin-labeled oligonucleotides can be purified from the unbiotinylated ones either by high-performance liquid chromatography *(10)*, polyacrylamide gel electrophoresis *(11)*, or by ion-exchange columns manufactured for this purpose (Perkin Elmer/ABI). If the biotin-labeled primer is not purified, the biotinylation should be confirmed after the PCR by affinity capture of the biotinylated PCR product followed by detection of possible unbound products by agarose gel electrophoresis, as follows: capture 10 µL of the PCR product

(or 1/10 of the reaction volume) on the avidin-coated beads in a total volume of 30 µL, as described in Section 3.2. Electrophorese the supernatant, containing the possible unbiotinylated PCR product, adjacent to 1/10 of the original PCR product, in an agarose gel containing ethidium bromide. If the biotinylation has been efficient, no product, or a faint product of significantly lower intensity than the unbound PCR product is observed in the supernatant.

2. The sequencing primer can be the unbiotinylated PCR primer or a nested primer. The use of a nested sequencing primer ensures that the possible nonspecific PCR products do not serve as templates in the sequencing reaction.

3. Wash the avidin-coated beads before use to eliminate the possible detached avidin as follows: mix the required amount of beads with 1 mL of the PBS/Tween buffer. Vortex well and centrifuge at 13,000g for 2 min. Discard the supernatant, and resuspend the pellet to the original volume with PBS/Tween buffer.

4. The avidin-coated beads are easy to handle when the solutions contain a detergent (e.g., 0.1% Tween 20) and the concentration of NaCl is not above 150 mM.

5. If the PCR product is long, the 3'-end of the captured single-stranded template may hybridize with homologous sequences within the template and serve as a sequencing primer. In our hands, this has not happened with PCR products shorter than 2 kb.

6. If the PCR product is to be purified from an agarose gel, purification protocols, including steps of phenol extraction, are not suitable for biotin-labeled PCR products. We have successfully used commercial glass-absorption purification kits for PCR products (e.g., Wizard PCR Preps, Promega Corp., Madison, WI).

7. The accurate molar amount of template DNA is not crucial in this protocol, but about 1 pmol of the template DNA is optimal.

8. The removal of all the supernatant is crucial to avoid an increase of the reaction volume in the following sequencing procedure.

9. When sequencing protocols other than the one presented here are applied, it should be noted that the volume of the template DNA captured on the beads is zero, since the particles are nonporous.

10. The protocol presented here can be used equally well with a [^{32}P]-labeled sequencing primer or with fluorescent labeling. Either a sequencing primer that has been fluorescently labeled at its 5'-end during the oligonucleotide synthesis or fluorescent dATP (Pharmacia Biotech) can be used, when the products of the sequencing reactions are analyzed using an automated DNA sequencer.

11. Sequence close to the primer (15–20 bp) is obtained by adding 1 µL of 0.1M MnCl$_2$ in 0.15M sodium isocitrate, into the mixture. The labeling mixture can also be diluted more, e.g., 1:8, and the labeling time can be reduced to 1–2 min to obtain sequence close to the primer.

12. Note that the beads can be loaded on the sequencing gel along with the sample.

References

1. Casanova, J.-L., Pannetier, C., Jaulin, C., and Kourilsky, P. (1990) Optimal conditions for directly sequencing double-stranded PCR products with Sequenase. *Nucleic Acids Res.* **18,** 4028.

2. Gyllensten, U. B. and Erlich, H. (1988) Generation of single stranded DNA by the polymerase chain reaction and its application to direct sequencing of the HLA-DQ alpha-locus. *Proc. Natl. Acad. Sci. USA* **85,** 7652–7656.

3. Stoflet, E. S., Koeberl, D. D., Sarkar, G., and Sommer, S. S. (1988) Genomic amplification with transcript sequencing. *Science* **239,** 491–494.

4. Syvänen, A.-C., Bengtström, M., Tenhunen, J., and Söderlund, H. (1988) Quantification of polymerase chain reaction products by affinity-based hybrid collection. *Nucleic Acids Res.* **16,** 11,327–11,338.

5. Syvänen, A.-C., Aalto-Setälä, K., Kontula, K., and Söderlund, H. (1989) Direct sequencing of affinity-captured amplified human DNA: application to the detection of apolipoprotein E polymorphism. *FEBS Lett.* **258,** 71–74.

6. Hultman, T., Ståhl, S., Hornes, E., and Uhlén, M. (1989) Direct solid phase sequencing of genomic and plasmid DNA using magnetic beads as solid support. *Nucleic Acids Res.* **17,** 4937–4946.

7. Hultman, T., Bergh, S., Moks, T., and Uhlén, M. (1991) Bidirectional solid-phase sequencing of in vitro amplified plasmid DNA. *BioTechniques* **10,** 84–93.

8. Lagerkvist, A., Stewart, J., Lagerström-Fermér, M., and Landegren, U. (1994). Manifold sequencing: efficient processing of large sets of sequencing reactions. *Proc. Natl. Acad. Sci. USA* **91,** 2245–2249.

9. Syvänen, A.-C. and Söderlund, H. (1993) Quantification of polymerase chain reaction products by affinity-based collection. *Methods Enzymol.* **218,** 474–490.

10. Bengtström M., Jungell-Nortamo, A., and Syvänen, A.-C. (1990) Biotinylation of oligonucleotides using a water soluble biotin ester. *Nucleosides Nucleotides* **9,** 123–127.

11. Wu, R., Wu, N.-H., Hanna, Z., Georges, F., and Narang, S. (1984) in *Oligonucleotide Synthesis: a Practical Approach* (Gait, M. J., ed.), IRL, Oxford, p. 135.

10

Analysis of Nucleotide Sequence Variations by Solid-Phase Minisequencing

Anu Suomalainen and Ann-Christine Syvänen

1. Introduction

The Sanger dideoxy-nucleotide sequencing method has been simplified by a number of methodological improvements, such as the use of the PCR technique for generating DNA templates in sufficient quantities for sequencing, the use of affinity-capture techniques for convenient and efficient purification of the PCR fragments for sequencing, the development of laboratory robots for carrying out the sequencing reactions, and the development of instruments for automatic on-line analysis of fluorescent products of the sequencing reactions. Despite these technical improvements, the requirement for gel electrophoretic separation remains an obstacle, when sequence analysis of large numbers of samples are needed, as in DNA diagnosis, or in the analysis of sequence variation for genetic, evolutionary, or epidemiological studies.

We have developed a method for analysis of DNA fragments differing from each other in one or a few nucleotide positions *(1)*, called solid-phase minisequencing, in which gel electrophoretic separation is avoided. Analogous to the methods for solid-phase sequencing of PCR products, the solid-phase minisequencing method is based on PCR amplification using one biotinylated and one unbiotinylated primer, followed by affinity capture of the biotinylated PCR product on an avidin- or streptavidin-coated solid support. The nucleotide at the variable site is detected in the immobilized DNA fragment by a primer extension reaction: a detection step primer that anneals immediately adjacent to the nucleotide to be analyzed is extended by a DNA polymerase with a single labeled nucleotide complementary to the nucleotide at the variable site (Fig. 1). The amount of the incorporated label is measured, and it serves as a specific indicator of the nucleotide present at the variable site.

From: *Methods in Molecular Biology, Vol. 65: PCR Sequencing Protocols*
Edited by: R. Rapley Humana Press Inc., Totowa, NJ

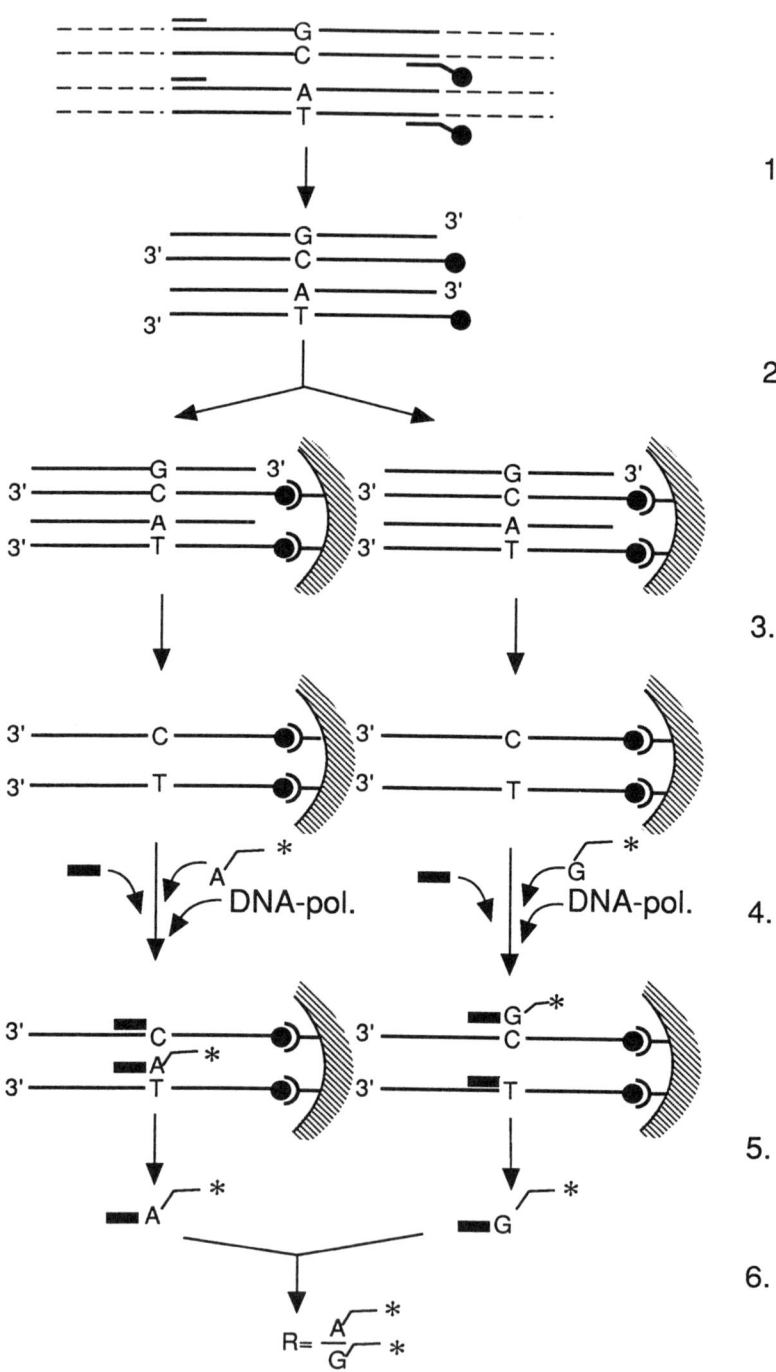

1.

2.

3.

4.

5.

6.

We have used the solid-phase minisequencing method for detecting numerous mutations causing human genetic disorders *(2)*, for analyzing allelic variation in genetic linkage studies, and for identification of individuals *(3)*. The protocol presented below is generally applicable for detecting any variable nucleotide. The method is well suited for analyzing large numbers of samples because it comprises simple manipulations in a microtiter plate or test tube format, and the result of the assay is obtained as an objective numeric value, which is easy to interpret. Furthermore, the solid-phase minisequencing method allows quantitative detection of a sequence variant present as a minority of <1% in a sample *(4)*. We have utilized the possibility of the sensitive quantitative analysis for detecting point mutations in malignant cells, present as a minority in a cell population *(4)* and for analyzing heteroplasmic mutations of mitochondrial DNA *(5,6)*. The high sensitivity is an advantage of the minisequencing method, compared to dideoxy-nucleotide sequencing, in which a sequence variant must be present as 10–20% of a mixed sample in order to be detectable. On the other hand, a limitation of the solid-phase minisequencing method is that it is restricted to analyzing variable nucleotides only at positions predefined by the detection step primers used.

2. Materials

2.1. Equipment and Materials

1. One of the PCR primers should be biotinylated at its 5'-end during the oligonucleotide synthesis, and the other primer is not biotinylated, resulting in a PCR product with one biotinylated strand *(see* Note 1).
2. Detection step primer: an oligonucleotide complementary to the biotinylated strand, designed to hybridize with its 3'-end with the nucleotide adjacent to the variant nucleotide to be analyzed *(see* Fig. 1 and Note 2).
3. Facilities for PCR.
4. Microtiter plates with streptavidin-coated wells (e.g., Combiplate 8, Labsystems, Finland) *(see* Note 3.)
5. Shaker at 37°C.
6. Water bath or incubator at 50°C.
7. Liquid scintillation counter.
8. Multichannel pipet and microtiter plate washer (optional).

Fig. 1. *(previous page)* Steps of the solid-phase minisequencing method. 1. PCR with one biotinylated (black ball) and one unbiotinylated primer. 2. Affinity capture of the biotinylated PCR product in streptavidin-coated microtiter wells. 3. Washing and denaturation. 4. The minisequencing primer extension reaction. 5. Measurement of the incorporated label. 6. Calculation of the result.

2.2. Reagents

All the reagents should be of standard molecular biology grade. Use sterile distilled or deionized water.

1. PBS/Tween solution: 20 m*M* sodium phosphate buffer, pH 7.5, 0.1% (v/v) Tween 20. Store at 4°C. Fifty milliliters are enough for several full-plate analyses.
2. TENT solution: 40 m*M* Tris-HCl, pH 8.8, 1 m*M* EDTA, 50 m*M* NaCl, 0.1% (v/v) Tween 20. Store at 4°C. Prepare 1–2 L at a time, which is enough for several full-plate analyses.
3. 50 m*M* NaOH (make fresh every 4 wk). Store at room temperature (about 20°C). Prepare 50 mL.
4. Thermostable DNA polymerase. We use *Thermus aquaticus* (*Taq*) DNA polymerase (Promega Corp., Madison, WI, 5U/µL) (*see* Note 4).
5. 10X concentrated *Taq* DNA polymerase buffer. We use: 500 m*M* Tris-HCl, pH 8.8, 150 m*M* $(NH_4)_2SO_4$, 15 m*M* $MgCl_2$, 1% v/v Triton X-100, 0.1% w/v gelatin. Store at –20°C.
6. [³H]-labeled deoxynucleotides (dNTPs): dATP to detect a T at the variant site, dCTP to detect a G, and so forth (Amersham, Buckinghamshire, UK [³H] dATP, TRK 625; dCTP, TRK 576; dGTP, TRK 627; dTTP, TRK 633), store at –20°C (*see* Note 5).
7. Scintillation reagent (for example Hi-Safe II, Wallac, Finland).

3. Method

3.1. PCR for Solid-Phase Minisequencing Analysis

Otherwise the PCR is done according to routine protocols, except that the amount of the biotin-labeled primer used should be reduced not to exceed the biotin-binding capacity of the microtiter well (*see* Note 3). For a 50-µL PCR reaction, we use 10 pmol of biotin-labeled primer and 50 pmol of the unbiotinylated primer. The PCR should be optimized (i.e., the annealing temperature and template amount) to be efficient and specific: to be able to use [³H] dNTPs, which are low-energy β-emitters, for the minisequencing analysis, 1/10 of the PCR product should produce a single visible band after agarose gel electrophoresis, stained with ethidium bromide. After optimization, there is no need for purification of the PCR product.

3.2. Solid-Phase Minisequencing Analysis

1. Affinity capture: Transfer 10-µL aliquots of the PCR product and 40 µL of the PBS/Tween solution to two streptavidin-coated microtiter wells (*see* Note 6). Include a control reaction, i.e., a well with no PCR product. Seal the wells with a sticker, and incubate the plate at 37°C for 1.5 h with gentle shaking.
2. Discard the liquid from the wells, and tap the wells dry against a tissue paper.
3. Wash the wells three times at room temperature as follows: pipet 200 µL of TENT solution to each well, discard the washing solution, and empty the wells thoroughly between the washings (*see* Note 7).

4. Denature the captured PCR product by adding 100 μL of 50 mM NaOH to each well, and incubate at room temperature for 3 min. Discard the NaOH, and wash the wells as in step 3.

5. For each DNA fragment to be analyzed, prepare two 50–μL mixtures of nucleotide-specific minisequencing solution, one for detection of the normal and one for the mutant nucleotide (*see* Note 8): Mix 5 μL of 10X *Taq* DNA polymerase buffer, 10 pmol of detection step primer, 0.2 μCi (usually equals to 0.2 μL) of one [^{3}H] dNTP, 0.1 U of *Taq* DNA polymerase, and dH$_2$O to a total volume of 50 μL It is obviously convenient to prepare master mixes for the desired number of analyses with a certain nucleotide.

6. Pipet 50 μL of one nucleotide-specific mixture per well. Incubate the plate at 50°C for 10 min in a water bath, or 20 min in an oven (*see* Note 9).

7. Discard the contents of the wells, and wash them as in step 3.

8. Release the detection step primer from the template by adding 60 μL 50 mM NaOH and incubating for 3 min at room temperature.

9. Transfer the eluted primer to the scintillation vials, add scintillation reagent, and measure the radioactivity, i.e., the amount of incorporated label, in a liquid scintillation counter (*see* Note 10).

10. The result is obtained as counts per minute (cpm) values. The cpm value of each reaction expresses the amount of the incorporated [^{3}H] dNTP. Calculate the ratio (R) between the mutant and normal nucleotide cpms. In a sample of a subject homozygous for the mutant nucleotide, the R will be >10, in a homozygote for the normal nucleotide, R will be <0.1, and in the case of a heterozygote, R varies between 0.5 and 2.0, depending on the specific activities of the [^{3}H] dNTPs (*see* Note 11).

4. Notes

1. The efficiency of the 5'-biotinylation of an oligonucleotide on a DNA synthesizer is most often 80–90%. The biotin-labeled oligonucleotides can be purified from the unbiotinylated ones either by high-performance liquid chromatography *(7)*, polyacrylamide gel electrophoresis *(8)*, or by ion-exchange columns manufactured for this purpose (Perkin Elmer/ABI, Foster City, CA). If the biotin-labeled primer is not purified, the biotinylation should be confirmed after the PCR by affinity capture of the biotinylated PCR product followed by detection of possible unbound products by agarose gel electrophoresis, as described in detail in Note 1 to Chapter 9 of this vol.

2. The detection step primer for our standard protocol is a 20 mer. It is advisable to use a nested primer as a detection step primer to ensure that possible unspecific PCR products remain undetected. The primer should be at least five nucleotides nested in relation to the unbiotinylated PCR primer.

3. The binding capacity of a streptavidin-coated microtiter well is 2–5 pmol of biotinylated oligonucleotide. If higher binding capacity is desired, avidin-coated polystyrene beads (Fluoricon, 0.99 μm, IDEXX Corp., Portland, ME; biotin-binding capacity over 2 nmol of oligonucleotide/mg beads) or streptavidin-coated magnetic polystyrene beads (Dynabeads M-280 Dynal A.S., Norway, streptavidin; biotin-binding capacity 300 pmol/mg) can be used *(9)*. The biotin-binding capac-

ity of a microtiter well allows reliable detection of up to 2% of a sequence variant present in the sample *(6)*, whereas a detection sensitivity of <0.1% is obtained with the bead-based format *(4)*.

4. It is advantageous to use a thermostable DNA polymerase for the single-nucleotide primer extension reaction, since a high temperature, favorable for the simultaneous primer annealing reaction, can be used.

5. Although the specific activities of the [^{3}H]NTPs are low, their half-lives are long (13 yr), and the necessary precautions for working with [^{3}H] should be taken. Also dNTPs or dideoxy-nucleotides labeled with other isotopes ([^{35}S] or [^{32}P]) or with haptens can be used *(1,10)*.

6. Each nucleotide to be detected at the variant site is analyzed in a separate well. Thus, at least two wells are needed per PCR product.

7. The washings can be performed utilizing an automated microtiter plate washer, or by manually pipeting the washing solution to the wells, discarding the liquid and tapping the plate against a tissue paper. Thorough emptying of the wells is important to avoid nonspecific nucleotide incorporation.

8. The minisequencing reaction mixture can be stored at room temperature for 1–2 h. It is convenient to prepare it during the incubation of Section 3.1., step 1.

9. The conditions for hybridizing the detection step primer are not stringent, and the temperature of 50°C can be applied to analysis of most PCR products irrespective of the sequence of the detection step primer. If the primer, however, is considerably shorter than a 20 mer or its GC content is low (melting temperature close to 50°C), lower temperatures for the primer annealing may be required.

10. Streptavidin-coated microtiter plates made of scintillating polystyrene are available (ScintiStrips, Wallac, Finland). When these plates are used, the final washing, denaturation, and transfer of the eluted detection primer can be omitted, but a scintillation counter for microtiter plates is needed *(11)*.

11. The ratio between the cpm values for the two nucleotides reflects the ratio between the two sequences in the original sample. Therefore, the solid-phase minisequencing method can be used for quantitative PCR analyses *(4–6)*. The *R*-value is affected by the specific activities of the [^{3}H] dNTPs used, and if either the mutant or the normal sequence allows the detection step primer to be extended by more than one [^{3}H] dNTP, this will obviously also affect the *R*-value. Both of these factors can easily be corrected when calculating the ratio between the two sequences. Another possibility is to construct a standard curve by mixing the two sequences in known ratios, and plotting the obtained cpm values as a function of the ratios to obtain a linear standard curve *(5,6)*. The test results can then be interpreted from the standard curve without the need of taking the specific activities of the number of [^{3}H] dNTPs incorporated into account.

References

1. Syvänen, A.-C., Aalto-Setälä, K., Harju, L., Kontula, K., and Söderlund, H. (1990) A primer-guided nucleotide incorporation assay in the genotyping of apolipoprotein E. *Genomics* **8**, 684–692.

2. Syvänen, A.-C. (1994) Detection of point mutations in human genes by the solid-phase minisequencing method. *Clin. Chim. Acta* **226,** 225–236.
3. Syvänen, A-C., Sajantila, A., and Lukka, M. (1993) Identification of individuals by analysis of biallelic DNA markers, using PCR and solid-phase minisequencing. *Am. J. Hum. Genet.* **52,** 46–59.
4. Syvänen, A.-C., Söderlund, H., Laaksonen, E., Bengtström, M., Turunen, M., and Palotie, A. (1992) N-ras gene mutations in acute myeloid leukemia: accurate detection by solid-phase minisequencing. *Int. J. Cancer* **50,** 713–718.
5. Suomalainen, A., Kollmann, P., Octave, J.-N., Söderlund, H., and Syvänen, A.-C. (1993) Quantification of mitochondrial DNA carrying the $tRNA_{8344}^{Lys}$ point mutation in myoclonus epilepsy and ragged-red-fiber disease. *Eur. J. Hum. Genet.* **1,** 88–95.
6. Suomalainen, A., Majander, A., Pihko, H., Peltonen, L., and Syvänen, A.-C. (1993) Quantification of $tRNA_{3243}^{Leu}$ point mutation of mitochondrial DNA in MELAS patients and its effects on mitochondrial transcription. *Hum. Mol. Genet.* **2,** 525–534.
7. Bengtström, M., Jungell-Nortamo, A., and Syvänen, A.-C. (1990) Biotinylation of oligonucleotides using a water soluble biotin ester. *Nucleosides Nucleotides* **9,** 123–127.
8. Wu, R., Wu, N.-H., Hanna, Z., Georges, F., and Narang, S. (1984) in *Oligonucleotide Synthesis: a Practical Approach* (Gait, M. J., ed.), IRL, Oxford, P. 135.
9. Syvänen, A.-C. and Söderlund, H. (1993) Quantification of polymerase chain reaction products by affinity-based collection. *Methods Enzymol.* **218,** 474–490.
10. Harju, L., Weber, T., Alexandrova, L., Lukin, M., Ranki, M., and Jalanko, A. (1993) Colorimetric solid-phase minisequencing assay illustrated by detection of alpha-l-antitrypsin Z mutation. *Clin. Chem.* **39,** 2282–2287.
11. Ihalainen, J., Siitari, H., Laine, S., Syvänen, A.-C., and Palotie, A. (1994) Towards automatic detection of point mutations: use of scintillating microplates in solid-phase minisequencing. *BioTechniques* **16,** 938–943.

11

Nonradioactive PCR Sequencing Using Digoxigenin

Siegfried Kösel, Christoph B. Lücking, Rupert Egensperger, and Manuel B. Graeber

1. Introduction

Techniques for direct sequencing of PCR products are of central importance to contemporary research in molecular biology and genetics. The rapidly growing number of cloned human disease genes increasingly allows sequencing of PCR amplicons for diagnostic purposes. Nonradioactive sequencing protocols are of particular use because health, environmental, and administrative risks are minimized compared with conventional isotopic methods. The PCR-based nonradioactive cycle sequencing protocol described in this chapter has been successfully employed to sequence mitochondrial and nuclear genes in Parkinson's and Alzheimer's disease brains using DNA extracted from formalin-fixed and paraffin-embedded neuropathological material (1–3). This method, which allows sequence information of PCR products to be obtained within a single day, can be carried out in a research or clinical laboratory using relatively inexpensive equipment.

Following initial PCR amplification, amplicons are purified using spin columns for affinity chromatography or ultrafiltration. Subsequently, cycle sequencing (4) is performed using 5'-digoxigenin end-labeled oligonucleotide primers. Because the nucleotide sequences of the PCR and sequencing primers can be identical, both reactions may be performed using the same thermal cycling protocol. This obviates the need for time-consuming optimization procedures. For visualization of sequencing results, sequencing reactions are separated on a standard sequencing gel, the gel is contact-blotted to a nylon membrane, and sequencing bands are visualized using alkaline phosphatase-conjugated antibodies (Fig. 1).

Potential pitfalls of our method are primarily related to the extreme sensitivity of PCR. The need for positive and negative sample controls cannot be over-

From: *Methods in Molecular Biology, Vol. 65: PCR Sequencing Protocols*
Edited by: R. Rapley Humana Press Inc., Totowa, NJ

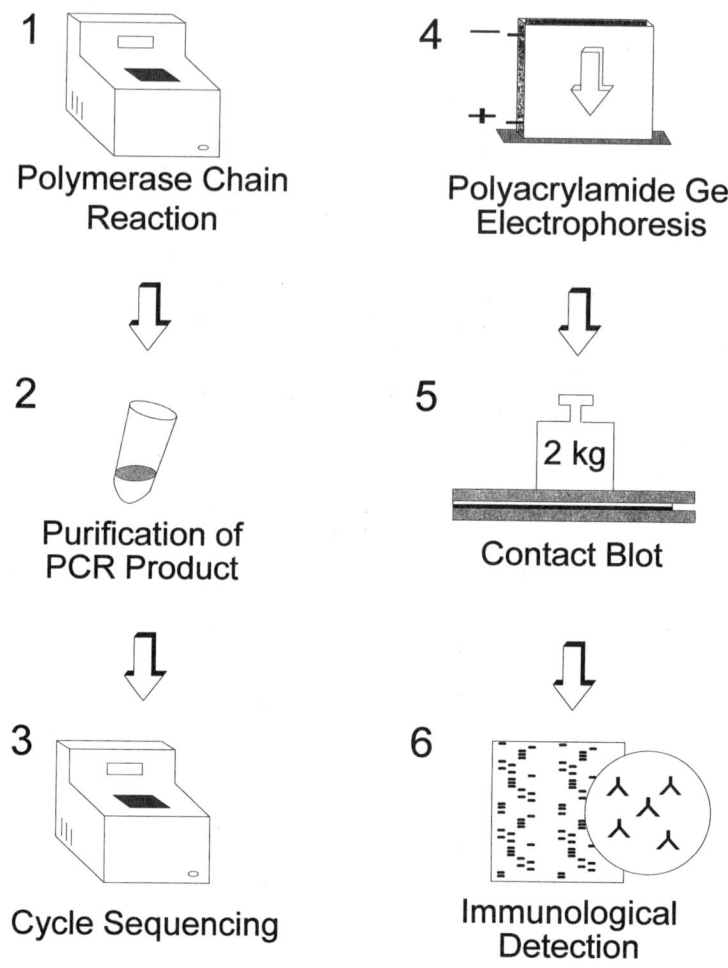

Fig. 1. Schematic drawing summarizing the essential steps of nonradioactive PCR sequencing using digoxigenin.

emphasized. We use different rooms and different pipets for setting up PCR reactions, pipeting sequencing templates, and thermal cycling *(5)*. In addition, aerosol-resistant pipet tips are always used.

2. Materials

2.1. Purification of Sequencing Templates

1. 10X TNE: 100 mM Tris-HCl, pH 7.4, 1.0M NaCl, 10 mM EDTA (*see* Note 1).
2. Wizard PCR Preps DNA Purification System containing affinity chromatography spin columns, purification buffer, and resin (A7170, Promega, Madison, WI).

Not contained in the kit are 1X TE: 10 mM Tris-HCl, pH 8.0, 1 mM EDTA *(6)*, 80% isopropanol, and 2-mL disposable syringes (one per reaction), or alternatively, Microcon-30 Concentrators (#42410, Amicon, Beverly, MA).
3. 1.5-mL Eppendorf tubes.
4. Eppendorf centrifuge (#5415C Eppendorf, Hamburg, Germany).
5. Fluorometer (TKO 100, Hoefer, San Francisco, CA) for quantification of template concentrations using DNA dye Hoechst No. 33258 and calf thymus standard DNA (supplied with the fluorometer).

2.2. Sequencing Reactions

1. Sequencing primer, 5'-digoxigenin end-labeled (1 pmol/μL), desalted, and HPLC-purified (*see* Note 2).
2. Mineral oil (M-5904, Sigma, St. Louis, MO).
3. Dig Taq DNA Sequencing Kit (#1449443, Boehringer Mannheim, Mannheim, Germany), containing:
   a. *Taq* DNA polymerase (3 U/μL).
   b. Sequencing buffer: 250 mM Tris-HCl, pH 8.0, 50 mM MgCl$_2$.
   c. Termination mixtures containing dNTPs and the appropriate ddNTP:
      ddATP (dATP, dCTP, dGTP, and dTTP, 25 μM each; 850 μM ddATP; 950 μM MgCl$_2$; pH 7.5).
      ddCTP (dATP, dCTP, dGTP, dTTP, 25 μM each; 400 μM ddCTP; 500 μM MgCl$_2$; pH 7.5).
      ddGTP (dATP, dCTP, dGTP, dTTP, 25 μM each; 75 μM ddGTP; 175 μM MgCl$_2$; pH 7.5).
      ddTTP (dATP, dCTP, dGTP, dTTP, 25 μM each; 1275 μM ddTTP; 1370 μM MgCl$_2$; pH 7.5).
      A second set of termination mixtures is shipped with the kit substituting 7-deaza-dGTP for dGTP.
   d. Loading buffer containing formamide.
4. 0.5-mL thin-walled reaction tubes (N801-0537, Applied Biosystems, Foster City, CA).
5. Eppendorf centrifuge.
6. Thermal cycler for cycle sequencing (*see* Note 6 for reference).

2.3. Preparation of Sequencing Gel

1. 10X TBE: 0.9M Tris-HCl, pH 8.3, 0.9M boric acid, and 25 mM EDTA *(6)*.
2. 10% Ammonium persulfate (w/v) (should be prepared freshly before use).
3. Sigmacote (SL-2, Sigma).
4. GelMix 8 (#5545UA, Gibco BRL, Gaithersburg, MD), containing 7.6% acrylamide (w/v), 0.4% *N,N*'-methylene bis-acrylamide, 7.0M urea, 100 mM Tris-borate, pH 8.3, 1 mM Na$_2$EDTA, 3 mM TEMED.
5. 10-mL disposable syringes with needles.
6. Scotch electrical tape 50 mm (FE-5000-0409-1, 3M).
7. Plastic foil (e.g., Saran Wrap™).

2.4. Gel Electrophoresis
and Visualization of Sequencing Results

1. 1X TBS: 50 mM Tris-HCl, pH 7.5, 150 mM NaCl.
2. Alkaline phosphatase reaction buffer: 0.1M Tris-HCl, pH 9.5, 50 mM MgCl$_2$, 0.1M NaCl.
3. Stock solutions of 70% (v/v) and 100% N,N-dimethyl formamide.
4. Digoxigenin Detection Kit for Glycoconjugate and Protein Analysis (#1210220, Boehringer Mannheim), containing:
   a. Antidigoxigenin antibodies conjugated to alkaline phosphatase.
   b. Blocking reagent (purified casein fraction).
   c. Nitroblue tetrazoliumchloride (NBT), 77 mg/mL in 70% N,N-dimethyl formamide.
   d. X-Phosphate (5-bromo-4-chloro-3-indolylphosphate, 4-toluidinium salt), 50 mg/mL in 100% N,N-dimethyl formamide.
5. Nylon membrane, positively charged (#1417240, Boehringer Mannheim).
6. Whatman chromatography paper (#3030917, 3MM Whatman International Ltd., Maidstone, UK).
7. Plastic hybridization bags.
8. Scalpel blade.
9. Standard sequencing equipment: sequencing apparatus (Model S2, Gibco BRL) and high-voltage power supply (PS 9009, Gibco BRL).
10. UV transilluminator (302 nm) for DNA crosslinking (Hoefer).
11. Eppendorf centrifuge.
12. Heating block or thermal cycler for denaturation of DNA.
13. Electric sealer for closing hybridization bags.

3. Methods
3.1. Purification of Sequencing Templates

1. PCR is performed according to established protocols *(7)*. Following PCR amplification, PCR products are purified away from excess nucleotides and primers using either spin column chromatography (Wizard PCR Preps DNA Purification System, Promega, Madison, WI) or ultrafiltration (Microcon-30, Amicon). The purification systems are used according to manufacturer's recommendation (*see* Note 3).
2. Measure DNA concentration using the fluorometer and Hoechst dye No. 33258. The dye is dissolved in 1X TNE (prepare two stocks, 0.1 and 1 µg/mL, respectively). Calf thymus DNA (100 or 1000 ng/µL) is used as a standard for calibration (*see* Note 4).
3. Purification results may be checked by electrophoresis of samples through a 2% high-melting-point agarose gel.

3.2. Sequencing Reactions

1. Sequencing reactions are set up in a total volume of 20 µL containing:
   a. 13 µL DNA template solution (2–4 pmol); dilute with sterile, double-distilled water, if necessary.

b. 3 µL 5'-digoxigenin end-labeled primer (1 pmol/µL).

c. 2 µL 10X reaction buffer (250 mM Tris-HCl, pH 8.0, 50 mM MgCl$_2$).

d. 2 µL *Taq* polymerase (3 U/µL).

2. For each sequencing reaction, transfer 4 µL of the above mixture to four thin-walled PCR reaction tubes, each containing 2 µL of the respective termination mixture (ddATP, ddCTP, ddTTP, and ddGTP) (*see* Note 5).

3. Overlay samples with 20 µL of mineral oil, and centrifuge for a few seconds in an Eppendorf tube, centrifuge at full speed.

4. Cycle sequencing is performed using a thermal cycling protocol empirically optimized for the T_m of the sequencing primer. When using sequencing primers of the same sequence and length as those used for PCR, a thermal cycling protocol identical to the one employed for PCR usually gives good results (*see* Note 6).

3.3. Preparation of Sequencing Gel

For gel electrophoretic separation of sequencing reactions, an 8 % denaturing polyacrylamide gel and standard sequencing equipment are used:

1. Clean glass plates thoroughly with 70% ethanol. Cover inner surface of one plate with a few drops or Sigmacote, and let evaporate for 5–10 min. Set up glass plates with spacers and seal edges airtight using Scotch electrical tape. Put two strong metal clamps on each side.

2. Add 450 µL of 10% ammonium persulphate to one bottle (75 mL) of GelMix 8, and mix gently. Slowly fill the space between the glass plates (avoid air bubbles!). Let gel polymerize for at least 1 h at room temperature. (Wear protective gloves when handling unpolymerized acrylamide.)

3. After polymerization of gel, remove clamps and electrical tape from the lower end of the glass plates, and transfer gel to sequencing apparatus. Fill the upper and lower buffer chambers with 500 mL of 1X TBE each. Rinse sample wells of gel with 1X TBE using a 10-mL disposable syringe with needle.

3.4. Gel Electrophoresis and Visualization of Sequencing Results

3.4.1. Electrophoresis and Contact Blotting

1. Add 3 µL of loading buffer to each tube containing the sequencing reactions with the respective "A," "C," "G," and "T" termination mixtures. Centrifuge for a few seconds to mix, and denature samples for at least 2 min at 85°C using a heating block or thermal cycler.

2. Transfer 6 µL of each sample to the wells of the sequencing gel. Run gel at 35 mA for 2–4 h (until the upper dye front has reached the bottom end of the gel).

3. Remove glass plates from electrophoresis apparatus. Carefully take off the glass plate covered with Sigmacote on the inner side using a scalpel blade as a lever. Cover gel with nylon membrane (avoid air bubbles!), and cover with one layer of Whatman filter paper. Put on second glass plate, and add approx 2 kg of weight. Leave for 30 min (*see* Note 7). Protect nylon membrane with one layer of plastic foil.

4. Crosslink nylon membrane (plastic wrapped, blotted side down) on UV light box at 302 nm for 3 min (*see* Note 8). The membrane may now be air-dried, sealed in a plastic bag (*see* Note 9), and stored at 4°C until further use.

3.4.2. Visualization of Sequencing Bands

Visualization of sequencing results is performed in plastic hybridization bags (one blot per bag) at room temperature using the Digoxigenin Detection Kit (Boehringer). After crosslinking, wet blots may be used immediately for immunological detection. The following recipe applies to a 4 × 4 lane gel (gently agitate membrane, except during the final step when incubating with substrate solution):

1. Incubate nylon membrane (in hybridization bag) for at least 30 min in approx 30 mL blocking solution (0.5 g/100 mL TBS) (*see* Note 10).
2. Wash blot three times in approx 100 mL TBS (10 min each).
3. Incubate membrane with 50 mL alkaline phosphatase-conjugated antidigoxigenin antibody for 1 h (50 µL antiserum/50 mL TBS).
4. Wash three times in TBS (20 min each).
5. Prepare 30 mL of substrate solution (mix immediately before use):
   150 µL NBT (77 mg/mL in 70% *N,N*-dimethyl formamide [v/v]).
   112 µL X-phosphate (50 mg/mL in *N,N*-dimethyl formamide).
   30 mL alkaline phosphatase buffer, pH 9.5.
   Thoroughly remove washing solution from membrane. Add substrate solution. During this incubation step, cover plastic bag with black light protector and do not agitate (*see* Note 11).
6. Stop developing process when faint sequencing bands are beginning to appear. Add approx 1 L of deionized water to hybridization bag. Cut bag open and carefully remove membrane. Allow membrane to air-dry on sheets of Whatman filter paper or paper towels. Store stained, dried membranes in the dark (*see* Note 12). An example of a stained membrane is shown in Fig. 2.

4. Notes

1. Molecular-biology-grade reagents are used for all experiments.
2. Oligonucleotide primers for PCR and sequencing may be designed by hand, but use of a computer program, such as Oligo (National Biosciences, Oslo, Norway), greatly facilitates this task. We have had good experience with primers ranging from 18 to 25 bp in length for both PCR and sequencing. The sequencing primer may be internally nested or identical in position and/or length to the primer used for PCR. Primers used in our laboratory were synthesized on an Applied Biosystems 394 DNA synthesizer at the Genzentrum of the University of Munich. Sequencing primers were custom-made and hapten-labeled using digoxigenin-3-*O*-methylcarbonylε-amino-caproic acid-*N*-hydroxy-succinimide ester and a 5'-oligopeptide linker (#1333054, Boehringer Mannheim).
3. Purification of PCR products using the Wizard PCR Preps DNA Purification System: Add 50–100 µL of PCR product to 100 µL of purification buffer in an

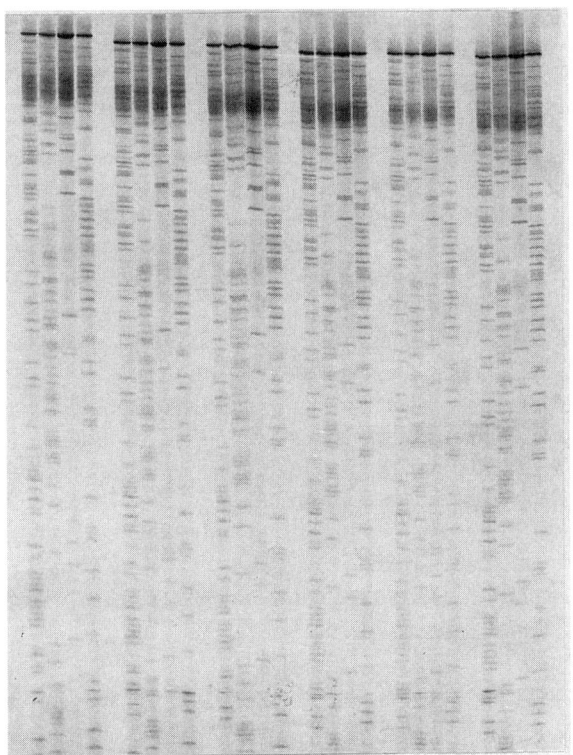

Fig. 2. Sequencing results obtained with primers *ND3* 5'-TCC CCA CCA TCA TAG CCA-3' and *ND4* 5'-GGG TTT TGC AGT CCT TAG-3', which amplify a 302-bp segment of the mitochondrial ND2 gene *(3)*. Sequencing results were visualized using alkaline phosphatase-conjugated antibodies directed against digoxigenin and NBT/X-phosphate as an enzymatic substrate. Contact blot of an 8% standard sequencing gel.

Eppendorf tube, and mix. Then add 1 mL of purification resin, and mix thoroughly. Pipet mixture into a 2-mL syringe, and push into a Wizard Prep column. Wash column with 2 mL of 80% isopropanol, and spin for 20 s in an Eppendorf centrifuge at full speed. Let isopropanol evaporate (takes approx 5 min), and transfer column to new Eppendorf tube. Elute DNA by adding 50 μL of 1X TE to the column. After 1 min, spin column for 20 s to recover DNA completely.

When ultrafiltration is used for purification of PCR products, 50–100 μL of PCR product are diluted with approx 400 μL double-distilled water and transferred to a Microcon-30 concentrator. Centrifugation of the Microcon-30 column is done according to manufacturer's recommendation (e.g., 10 min at 12,000g) in an Eppendorf centrifuge.

Important: Avoid contamination of both types of columns with mineral oil used for overlaying PCR reactions.

4. A fluorometer is used to measure nanogram amounts of DNA. The TKO 100 (Hoefer) is a useful and reasonably priced alternative to full-sized spectrophotometers for measuring concentrations of double-stranded DNA. DNA concentrations as low as 10 ng/μL may be reliably determined using a standard setup of the TKO. For measuring the concentration of PCR products, the fluorometer is first calibrated with DNA standard (100 or 1000 ng/μL calf thymus DNA, depending on the DNA concentrations to be determined): 1 μL of standard DNA is mixed with 1 mL of 1X TNE containing Hoechst dye No. 33258 and added to the quartz cuvet. Concentration of the Hoechst dye is 0.1 or 1 μL/mL in 1X TNE, depending on the concentration of the calibration standard.

5. Termination mixtures containing 7-deaza-dGTP are used to avoid band compression artifacts when sequencing GC-rich regions.

6. Similar to PCR, the success of *Taq* cycle sequencing depends on buffer conditions, especially the concentration of Mg^{2+} ions and pH. Buffer conditions for PCR and cycle sequencing may be optimized using commercial "optimizer" kits (e.g., K1220-01, Invitrogen, Leek, Netherlands). We have used *Taq* polymerases from Perkin-Elmer and Boehringer Mannheim with comparable success for both PCR and sequencing. Thermal cyclers from Perkin-Elmer (TC 480) and Biometra (Trio Thermoblock) also gave comparable results in our hands. Sequencing protocols employing smaller numbers of cycles, e.g., 20–25 may reduce the intensity of shadow bands.

7. Do not extend blotting time, and do not use higher weights.

8. Protect eyes and skin against UV light (wear goggles, mask, coat, and gloves)!

9. In order to save on reagents, hybridization bags should be tightly sealed. However, leave one "long end," because bags will need to be reopened and resealed during subsequent incubation steps.

10. Blocking reagent (Boehringer Mannheim) should be prepared freshly before use. Dissolve blocking reagent at 50°C in TBS, and let cool down to room temperature. Blots can be stored in blocking solution at 4°C for a few days.

11. To avoid diffuse or spotty background, the volume of the substrate solution should be sufficiently large to cover the nylon membrane completely. Also, avoid folds in the hybridization bag. Sequencing bands should become visible within 15 min on incubation. Do not move the membrane during that time. Since the visualization process involves an enzymatic reaction, lowering the temperature of the incubation solutions may lengthen incubation times. We recommend developing at room temperature and checking the intensity of the stained sequencing bands by briefly lifting the dark cover from time to time.

12. The intensity of the sequencing bands may increase during the first few hours after the final incubation has been finished (if alkaline phosphatase and NBT/X-phosphate are used for visualization as in the present protocol). Therefore, the incubation with substrate solution should be stopped as soon as the first sequencing bands are clearly readable.

    Important: Stained, dried nylon membranes should not be exposed to sunlight, since they will bleach rapidly. However, stained membranes can be stored safely for many months in the dark. For long-term documentation, we recommend taking photographs. Sequencing results may also be documented permanently by photocopying stained blots onto paper or overhead transparencies.

References

1. Kösel, S. and Graeber, M. B. (1993) Non-radioactive direct sequencing of PCR products amplified from neuropathological specimens. *Brain Pathol.* **3,** 421–424.
2. Kösel, S. and Graeber, M. B. (1994) Use of neuropathological tissue for molecular genetic studies: parameters affecting DNA extraction and polymerase chain reaction. *Acta Neuropathol.* **88,** 19–25.
3. Kösel, S., Egensperger, R., Mehraein, P., and Graeber, M. B. (1994) No association of mutations at nucleotide 5460 of mitochondrial NADH dehydrogenase with Alzheimer's disease. *Biochem. Biophys. Res. Commun.* **203,** 745–749.
4. Krishnan, B. R., Blakesley, R. W., and Berg, D. E. (1991) Linear amplification DNA sequencing directly from single phage plaques and bacterial colonies. *Nucleic Acids Res.* **19,** 1153
5. Kwok, S. and Higuchi, R. (1989) Avoiding false positives with PCR. *Nature* **339,** 237–238.
6. Sambrook, J., Fritsch, E. F., and Maniatis, T. (1989) *Molecular Cloning: A Laboratory Manual*, 2nd ed., Cold Spring Harbor Laboratory Press, Cold Spring Harbor, New York.
7. Saiki, R. K., Gelfand, D. H., Stoffel, S., Scharf, S. J., Higuchi, R., Horn, G. T., Mullis, K. B., and Erlich, H. A. (1988) Primer-directed enzymatic amplification of DNA with a thermostable DNA polymerase. *Science* **239,** 487–491.

12

Silver Sequencing™

Nonradioactive Cycle Sequencing of dsDNA

Alison Wade-Evans

1. Introduction

The conventional methods of sequencing, such as Maxam and Gilbert *(1)* (involving chemical cleavage of labeled DNA fragments) and Sanger et al. *(2)*, of dideoxy sequencing (using enzymatic extension of oligonucleotide primers) remain the most reliable techniques, but when compared to more modern methods, they are most labor-intensive. For rapid sequence analysis of limited quantities of template, cycle sequencing *(3)* has become increasingly popular. The technology, described in detail by Blakesley *(4)*, is basically as other dideoxy sequencing techniques, but using a thermostable polymerase, *Taq* polymerase *(5)*, which is resistant to multiple heat denaturation steps. This allows repriming from the same template, resulting in a linear increase in the quantity of synthesized product.

Until fairly recently, the above-mentioned techniques have involved the use of radioactive material, either incorporated into DNA fragments (Maxam and Gilbert *[1]*) or oligonucleotide primers, or as labeled nucleotides incorporated into the synthesized product. Because of the hazardous nature of radioactivity, alternative labeling methods have been extensively investigated. These alternatives include the following "tags": fluorescence, chemiluminescence, biotin, and digoxygenin, the use of which have been described in detail elsewhere in this volume. The main drawbacks to these approaches are financial in that either specialized equipment is necessary for analyzing the data and/or suitably modified ("tagged") nucleotides are essential. The technique described in this chapter, Silver Sequencing™, does not require the use of any specialized equipment

From: *Methods in Molecular Biology, Vol 65: PCR Sequencing Protocols*
Edited by: R. Rapley Humana Press Inc., Totowa, NJ

or modified nucleotides, thus cutting down the costs often associated with the other nonradioactive sequencing methods.

The Silver Sequencing system* combines the advantages of thermal cycle sequencing with a sensitive silver-staining protocol to visualize DNA bands separated on a normal 6% urea/acrylamide gel *(6)*. The *Taq* polymerase** used in this system is a modified form that has been patented by Promega, Madison, WI (US Patent No. 5,108,892), which gives superior results on double-stranded DNA templates (*see* Note 1). 7-Deaza dGTP (Boehringer-Mannheim, Indianapolis, IN, US Patent No. 4,804,748) has been substituted for dGTP in this system to assist in resolving band compressions associated with GC-rich regions One of the major differences between silver staining and other methods of sequencing is that in order to be able to stain the gel following separation of the extension products by electrophoresis, it must be fixed firmly to one of the glass plates. The long glass plate must be siliconized prior to pouring the gel and the short one treated with bind silane to ensure that the gel adheres to only one of the plates. Care must be taken to avoid crosscontamination of the plates with either compound. Careful plate preparation is essential for success. The gel plates are then assembled, and a 6% urea/acrylamide gel is poured and run as normal. Prior to staining, the sequencing gel is fixed in acetic acid to remove electrophoresis buffer and urea, and also to prevent diffusion of the smaller extension products. The gel is then rinsed thoroughly to remove excess acetic acid and urea/Tris-borate. The gel is stained in a solution containing silver nitrate and formaldehyde, rinsed very briefly to remove any excess silver, and then immediately developed in an alkaline sodium carbonate solution (containing formaldehyde and sodium sulfate). Under these conditions, it is postulated that the silver ions are reduced rapidly by the formaldehyde to metallic silver *(7)*, which it has been proposed then binds to the endocyclic ring nitrogen atoms present in the nucleotide bases of the DNA *(8)*. The reduction is catalyzed by metallic silver localized at the nucleation sites formed during the staining step. The reaction rate is temperature-dependent and can be controlled more easily at low temperatures (10°C). The reaction can be stopped by the addition of acetic acid, since silver ion reduction is considerably slower under acidic conditions. A final rinse in water removes all the excess reagents. After staining, the DNA sequence can be viewed directly with the aid of a light box or may be transferred to electrophoresis duplicating film (EDF) to produce a permanent copy, but this is optional. Unlike other methods of sequencing, the results can be read (approx 250–375 bases) 90 min after electrophoresis is complete.

*Patent pending.
**This product has not been licensed for use in the polymerase chain reaction (PCR) for amplifying nucleic acid covered by US Patent Nos. 4,683,195 and 4,683,202 assigned to Hoffmann-La Roche.

Table 1
Working Solutions for Nucleotide Mixes

Component	G mix	A mix	T mix	C mix
ddGTP	45 μ*M*	—	—	—
ddATP	—	525 μ*M*	—	—
ddTTP	—	—	900 μ*M*	—
ddCTP	—	—	—	300 μ*M*
7-deaza-dGTP	30 μ*M*	30 μ*M*	30 μ*M*	30 μ*M*
dATP	30 μ*M*	30 μ*M*	30 μ*M*	30 μ*M*
TTP	30 μ*M*	30 μ*M*	30 μ*M*	30 μ*M*
dCTP	30 μ*M*	30 μ*M*	30 μ*M*	30 μ*M*

2. Materials

2.1. Sequencing Reagents

All sequencing components should be stored at $-20°C$ and kept on ice during use. A kit is available from Promega (Madison, WI) that contains all the reagents necessary for cycle sequencing.

1. Sequencing-grade *Taq* polymerase (*see* Note 1).
2. 10X TE buffer: 100 m*M* Tris-HCl, pH 7.5 (at 25°C), 1 m*M* EDTA. Filter buffer to sterilize.
3. 5X Sequencing buffer: 250 m*M* Tris-HCl, pH 9.0 (at 25°C), 10 m*M* MgCl$_2$.
4. d/ddNTP mixes: Make 0.3 m*M* stock solutions of dATP, dCTP, 7-deaza-dGTP, and TTP in 1X TE buffer (diluted from 10X TE stock solution with ultrapure water). Store aliquots at $-20°C$. Make 10-m*M* stock solutions of ddATP, ddCTP, ddGTP, and ddTTP in 1X TE buffer. Store aliquots at $-20°C$. Make 200 μL working solutions of nucleotide mixes from dNTP and ddNTP stock solutions according to concentrations in Table 1.
5. Stop solution: 10 m*M* NaOH, 95% formamide, 0.05% bromophenol blue, and 0.05% xylene cyanol. Handle this solution with care, since formamide is a potential teratogen.
6. pGEM 3ZF(+): or any control DNA at a concentration of 1 μg/μL (*see* Note 2): available commercially.
7. pUC/M13 forward primer (24 mer) or other suitable primers: also available commercially from many suppliers. In general, longer primers (>24 bases) and primers having a high GC content (>50%) tend to give stronger signals.

2.2. Electrophoresis Reagents

1. 6% Urea/acrylamide solution can be purchased commercially (although premixed solutions have a limited half-life) or made directly from chemicals in the laboratory (19:1 acrylamide:bis-acrylamide ratio), but a fume hood must be used owing to the extremely toxic nature of acrylamide in a crystalline form. Acrylamide solutions should be stored in the dark.

2. 5*M* NaCl, made with ultrapure water.
3. 10X TBE: Tris/borate buffer used for running sequencing gels, 108 g Tris base, 55 g boric acid, and 9.3 g Na$_2$EDTA made up to 1 L with ultrapure water.
4. Bind silane: Used to bind the acrylamide gel to the glass plate. It is available commercially (Sigma, St. Louis, MO), but is toxic so should be used in a fume hood and handled with gloves.
5. SigmaCote™ or RepelCote™ are both available commercially, but are toxic so they should be used in a fume hood and handled with gloves.
6. 10% Acetic acid: Should be made fresh just prior to use in the purest water available, e.g., Milli-Q.

2.3. Staining Reagents

All staining reagents should be stored at room temperature and are stable for approx 6 mo. These reagents can also be purchased in a kit form (Promega).

1. Stain solution: Dissolve 1 g of silver nitrate in 1 L ultrapure water. This compound is highly toxic in powder form, so suitable precautions should be taken. Just before use, add 1.5 mL/L of 37% formaldehyde (available commercially). Handle with care, since formaldehyde is a possible carcinogen. Waste silver can be recovered from the used staining solution for recycling. Precipitate the silver by adding 5 g/L NaCl. Collect AgCl precipitate by filtration or allow to settle by gravity.
2. Developing solution: Make up just prior to use. Dissolve 30 g of anhydrous sodium carbonate (AR-grade) in 1 L of ultrapure water. Chill to 10°C in an ice bath. **Immediately** (*see* Note 3) before use, add 1.5 mL of 37% formaldehyde and 200 µL of 10 mg/mL sodium thiosulfate.
3. EDF film is available from Promega or Kodak, Rochester, NY.

2.4. Equipment

1. Thermal cycler, preferably not regulated by a Peltier element (prone to drifting after prolonged use).
2. Orbital shaker.
3. Vertical gel electrophoresis equipment suitable for DNA sequencing, preferably to take gel plates of 40 × 20 cm.
4. Three high-density polyethylene trays with dimensions slightly larger than short glass plate, e.g., 38 × 20 cm.

3. Methods
3.1. Sequencing Protocol

1. Prepare four 0.5-mL Eppendorf tubes (labeled A, C, G, or T) for each set of sequencing reactions by adding 2 µL of the appropriate d/ddNTP mix to the bottom of each tube.
2. For the cycle sequencing reaction, prepare a template mix for each plasmid. The mix contains 5–6 pmol of primer (*see* Note 4), 0.5–1 pmol of supercoiled dsDNA (*see* Note 5), and 5 µL 5X sequencing buffer in a total volume of 16 µL using ultrapure water.

3. Add 1 µL (5 U) of *Taq* polymerase, sequencing grade (*see* Note 1), to each template mix from step 2, and mix thoroughly by pipeting.
4. Add 4 µL of the mix from step 3 to the side of each of the four tubes (prepared in step 1) containing the d/ddNTP mixes.
5. Add a drop of mineral oil (approx 100 µL) to each tube.
6. Pulse-spin the tubes on a bench microcentrifuge.
7. Transfer the tubes to a Thermal Cycler preheated to 95°C (*see* Note 6), and incubate for 2 min at this temperature.
8. Cycle 60 times at 95°C for 45 s (denaturing), 42°C for 45 s (annealing), and 70°C for 60 s (extension) (*see* Note 7).
9. Samples may be stored at –70°C if they are not to be used immediately or at 4°C overnight.

3.2. Electrophoresis Conditions

1. The glass plates must be meticulously clean. Clean the plates with warm water and detergent. Rinse thoroughly with deionized water to remove any detergent residues (*see* Note 8). Give the plates a final wash in ethanol.
2. Short glass plate preparation (*see* Note 9): Prepare a fresh binding solution just before use by adding 3 µL bind silane to 1 mL of 95% ethanol, and 0.5% glacial acetic acid. Use this to coat a scrupulously clean short glass plate. After 4–5 min, gently wipe (*see* Note 10) the plate with a paper tissue saturated with 95% ethanol. Repeat this wash three times, using a fresh paper tissue each time (*see* Note 11).
3. Long glass plate preparation (*see* Note 9). Change gloves before preparing the long glass plate to prevent crosscontamination with the binding solution. Wipe a scrupulously clean long glass plate with a tissue saturated with SigmaCote solution. After 5–10 min, remove any excess solution with a tissue (*see* Note 12).
4. Prepare a 4–6% gel with 0.4-mm spacers (*see* Note 13). Do not tape the two gel plates together, and leave them overnight before pouring the gel, since crosscontamination occurs, which results in a torn gel.
5. Prerun a 6% urea/acrylamide gel (0.4 mm thick, 40 × 20 cm) for 30 min at a constant 35 W (maximum 1700 V).
6. Add 3 µL of stop solution to each tube of the sequencing reactions.
7. Heat to 70°C for 2 min.
8. Load 3–4 µL from each tube, carefully avoiding the mineral oil (*see* Note 14).
9. Run the gel at 35 W, until the bromophenol blue dye has migrated one-third of the way down the length of the gel plates. Add 1M final concentration of sodium acetate to the bottom reservoir (*see* Note 15).
10. Continue running the gel until the bromophenol blue dye just migrates off the end of the gel into the bottom reservoir.
11. Carefully pry the two plates apart, using a plastic wedge. The gel should be firmly fixed to the short glass plate (*see* Note 16).
12. Develop the gel on the glass plate in a photographic staining tray using the following procedure.

3.3. Staining Procedure

N.B.: The gel must be agitated at all times.

1. Place the gel in a shallow plastic tray, cover with 1 L of 10% glacial acetic acid, and agitate well for at least 20 min (the gel can be left overnight at this stage—*see* Note 17). This step removes electrophoresis buffer and urea from the gel, which can cause high background staining. It also prevents diffusion of small extension products out of the gel. If the developing solution has not already been chilled to 10°C, place on ice.
2. Wash three times (2 min each) with 1 L of ultrapure water to remove excess acetic acid (*see* Note 18).
3. Transfer the gel into 1 L of staining solution, and agitate well for 30 min. Close to the end of this stage, add 1.5 mL of 37% formaldehyde and 200 µL of 10 mg/mL sodium thiosulfate to the prechilled developing solution in a fresh tray (*see* Note 19.
4. Rinse the surface of the gel (to remove excess silver) by submerging the plate **briefly** in ultrapure water in a different tray (no longer than 5 s in total—*see* Note 20).
5. Transfer the plate immediately to a white tray (*see* Note 21) containing the developing solution cooled to 10°C (*see* Note 22).
6. When the bands reach the required intensity and the background is still low, fix the gel by adding 1 L of 10% acetic acid (*see* Note 17). Watch carefully at this stage to ensure that the reaction is stopped before the background becomes too intense (*see* Note 23).
7. Fix for 5–10 min.
8. Wash the gel twice in fresh water (*see* Note 24), and leave to dry on the bench at room temperature or by convection heating (*see* Note 25).
9. The sequencing data can be read directly from the gel with the aid of a light box.
10. A permanent copy of the gel can be produced using EDF. The use of this film increases the contrast of the bands, which can improve the readability of some gels. The gel must be completely dry before exposure to the EDF film.

3.4. EDF Film Development

1. In a dark room, place the dry, stained gel fixed to the glass plate (gel side uppermost) on a fluorescent light box. Overhead fluorescent lights may be used as an alternative.
2. With the aid of a safety light, place the film on top of the gel, oriented with the notch at the top left (*see* Note 26).
3. Place a clean, dry glass plate on top of the film.
4. Switch on the light box for approx 20 s. The exposure time can be optimized by exposing strips of the EDF film for various time periods (*see* Note 27). Twenty to 40 s generally gives good results. When using overhead lights, exposure times of 45–90 s are recommended. Shorter exposure periods result in a darker image, whereas a longer exposure may help to eliminate background.

5. Manually develop the film (*see* Note 28), or use a flat-bed automated processor. The following procedure is recommended:
   1–5 min in Kodak GBX developer.
   1-min wash with distilled water.
   3 min in Kodak GBX fixer.
   Finally, a 1-min wash in distilled water.

4. Notes

Many factors affect the quality of the sequence data obtained using this system. Some are related to the sequencing reactions, and others to the staining procedure.

1. The sequencing-grade *Taq* polymerase is a thermostable enzyme that replicates at an optimum temperature of 70°C. It is purified from *Thermus aquaticus* (strain YT1) and has been modified according to procedures developed by Promega (US Patent No. 5,108,892). It produces uniform band intensity, low background, and has good proofreading activity.
2. The DNA template concentration must be estimated by agarose gel electrophoresis (run alongside known standards) or fluorescent detection methods, since spectrophotometric methods can give falsely high estimates of concentration owing to contaminating chromosomal DNA, protein, RNA, ribonucleosides liberated by RNase digestion, and chemicals, which increase absorbance values at 260 nm.
3. Formaldehyde is inactivated rapidly at 4°C.
4. Primer concentration can be estimated using the following equation:

$$1 \text{ pmol} = (n/3) \text{ ng} \tag{1}$$

   where *n* is the number of bases in the primer.
5. Recommended amounts of template DNA depend on the template type. The template concentration is crucial to the success of the sequencing reaction. For PCR† products 60–120 fmol is recommended, for supercoiled plasmid 1–2 pmol and for λ or cosmid DNA, 31 fmol. The following general formulae apply:
   For dsDNA:

$$1 \text{ pmol} = n \times (6.6 \times 10^{-4}) \mu g \tag{2}$$

   where *n* is the number of base pairs.
   For ssDNA:

$$1 \text{ pmol} = n \times (3.3 \times 10^{-4}) \mu g \tag{3}$$

   where *n* is the number of bases.
      Supercoiled plasmid yields a weaker signal than relaxed dsDNA; therefore, more template is recommended.

---

†The PCR process for amplifying nucleic acid is covered by US Patent Nos. 4,683,195 and 4,683,202, assigned to Hoffmann-La Roche. Patents pending in other countries.

6. It is important to preheat the thermal cycler to 95°C to prevent nonspecifically annealed primers from being extended by the *Taq* polymerase.

7. The annealing temperature needs to be optimized for each primer/template combination and should, as a general rule, be 5–10°C lower than the T_M of the primer, where:

$$T_M = ([2 \times A/T] + [4 \times G/C])°C \qquad (4)$$

High annealing temperatures inhibit strand reannealing, reduce template secondary structure, and improve the stringency of primer hybridization. Therefore, the highest possible annealing temperature should be used. As a starting point, the following programs can generally be used. Ramp times should be as rapid as possible.

For a primer <24 bases or with a GC content <50%, 45–60 cycles at 95°C for 45 s, 42°C for 45 s, and 70°C for 1 min.

For a primer >24 bases or with a GC content >50%, 45–60 cycles at 95°C for 45 s and 70°C for 45 s (annealing and extension step).

8. Detergent microfilms left on the glass plates may cause a high background brown color, when the gel is stained.

9. The plates must be treated every time they are used. This is essential to prevent tearing of the gel during the silver-staining protocol. As the gel is chemically crosslinked to the glass plate, it is possible to use 4% urea/acrylamide gels to maximize the length of read.

10. Vigorous rubbing will remove an excessive amount of the bind silane, and the gel may not adhere well.

11. This is essential to prevent the binding solution from contaminating the long glass plate, which could result in a torn gel.

12. Excess SigmaCote may cause inhibition of staining.

13. Gels thinner than 0.4 mm may give weak signals, whereas gels thicker than 0.4 mm or higher than 6% may crack during drying. Do not use wedge gels, since these stain unevenly. If a wedge gel effect is required, *see* Note 15.

14. It is not necessary to remove the mineral oil overlay, but care must be taken to draw up only the blue sample from below the mineral oil when pipeting. Duck bill tips are ideal for this. If oil is loaded together with the sample, then it can cause disturbance of the sample and can result in it being forced out of the well and consequently lost.

15. The ionic strength of the bottom of the gel is increased simply by increasing the salt concentration in the bottom buffer chamber. This results in an electrolyte gradient. During the run, the salt electrophoreses into the gel, generating a reproducible and effective gradient. If <400 bases of sequence are required, the sodium acetate is added to the bottom reservoir (to a final concentration of 1M) at the beginning of electrophoresis.

16. If the gel tears because of sticking to both plates or does not adhere to either plate, then treatment of the plates was not performed carefully enough. This is an absolutely critical stage in the preparation.

17. Save the fix solution from this step to terminate the developing reaction in Section 3.3., step 6.
18. The water used for the staining procedure must be ultrapure or double-distilled. If the water contains contaminants, especially halogen and metal ions, the sequencing signal may not develop or only bands in the top half of the gel may be visible.
19. Always use separate trays for the stain and developer to prevent crosscontamination between these two solutions.
20. The length of the rinse step following staining is critical. The time taken to rinse the gel **and** transfer it into the developing solution must not take longer than 5–10 s. If the gel is rinsed for too long, the silver will be removed from the DNA and little or no signal will develop. If the rinse does proceed for too long, the gel can be restained at this stage.
21. Use a white tray for developing the gel, since it is easier to visualize the bands and thus terminate the development step at the correct time.
22. The developing solution must be prechilled to 10°C to minimize any background staining. The rate of development is highly temperature-sensitive.
23. The developed bands appear light, but the sequence ladder will darken on drying and subsequent exposure to EDF film. Prolonged development times result in a high background. It is better to stop development early than to overdevelop the gel.
24. Handle the gel with gloves at the edges to prevent fingerprints from appearing on the gel. Without a final rinse, the developing reagents dry on the gel surface, resulting in crystallization of the sodium carbonate (a white powder).
25. As mentioned in Note 23, the sequence ladder darkens on drying and makes the data easier to read. The dried gel can be removed from the glass plate by soaking in 10% sodium hydroxide for about 1 h.
26. EDF film is one-sided, and it is therefore important to ensure that the notched corner is positioned top left.
27. The optimal exposure time will vary depending on the light source.
28. Owing to the flexibility of this particular film, it is not possible to develop it in an automated processor that uses rollers arranged vertically. Only flat-bed rollers are suitable. The smaller automated processors are organized in this way.

References

1. Maxam, A. M. and Gilbert, W. (1977) A new method for sequencing DNA. *Proc. Natl. Acad. Sci. USA* **74,** 560–564.
2. Sanger, F., Nicklen, S., and Coulson, A. R. (1977) DNA sequencing with chain-terminating inhibitors. *Proc. Natl. Acad. Sci. USA* **74,** 5463–5467.
3. Murray, V. (1989) Improved double-stranded DNA sequencing using linear polymerase chain reaction. *Nucleic Acids Res.* **17,** 8889
4. Blakesley, R. W. (1993) Cycle sequencing, in *DNA Sequencing Protocols , vol. 23, Methods in Molecular Biology* (Griffin, H. G. and Griffin, A. M., eds.), Humana, Totowa, NJ, pp. 209–217.

5. Saiki, R. K., Gelfand, D. H., Stoffel, S., Scharf, S. J., Higuchi, R., Horn, G. T., Mullis, K. B., and Erlich, H. A. (1988) Primer-directed enzymatic amplification of DNA with a thermostable DNA polymerase. *Science* **239**, 487–491.
6. Bassam, B. J., Coetano-Annolles, G., and Gresshoff, P. M. (1991) Fast and sensitive silver staining of DNA in polyacrylamide gels. *Anal. Biochem.* **196**, 80–83.
7. Caetano-Anolles, G. and Gresshoff, P. M. (1994) DNA amplification fingerprinting using very short arbitrary oligonucleotide primers. *Promega Notes* **45**, 13.
8. Menzer, S., Sabat, M., and Lippert, B. (1992) AG(I) modified base pairs involving complementary (G,C) and noncomplementary (A,C) nucleobases—on the possible structural role of aqua ligands in metal-modified nucleobase pairs. *J. Am. Chem. Soc.* **114**, 4644–4649.

13

Direct Sequencing
of PCR Products with DNA-Binding Proteins

Ralph Rapley

1. Introduction

The polymerase chain reaction (PCR) has become widely established as a powerful core molecular biology technique because of its ability of produce large amounts of specific target DNA from limited template sources *(1)*. Numerous applications based on the PCR have also been developed, including site-directed mutagenesis, in vitro expression, and nucleotide sequencing *(2,3)*. The combination of PCR amplification and direct sequencing of products is a highly desirable procedure that, in many cases, obviates the need for complex and time consuming cloning protocols *(4)*. However, there appears to be no one universally accepted method for direct PCR sequencing *(5)*.

In order to analyze PCR amplified products of immunoglobulin-variable regions by nucleotide sequencing, we have incorporated single-stranded DNA-binding protein (SSB) from *Escherichia coli* in a direct sequencing strategy using the dideoxy chain termination method *(6)*. This may be undertaken using a Sequenase kit modified to include dimethylsulfoxide as previously described *(7)*. The mode of action of single-stranded DNA binding protein (SSB) in combination with the denaturant DMSO is probably through the prevention of complementary strand reassociation following the annealing of the sequencing primer. It is also possible that their inclusion may also prevent or reduce the formation of any DNA secondary structure, a particular problem that is known to cause polymerase pausing and premature termination in sequencing reactions. The use of DNA-binding proteins in amplification and sequencing may prove to be generally applicable in improving the yield and quality of a number of templates from various sources.

From: *Methods in Molecular Biology, Vol. 65: PCR Sequencing Protocols*
Edited by: R. Rapley Humana Press Inc., Totowa, NJ

2. Materials (*see* Note 1)

2.1. PCR Purification Reagents

1. TBE buffer: 890 mM Tris-borate, 890 mM boric acid, 20 mM EDTA, pH 8.3.
2. Water-equilibrated diethyl ether.
3. Butan-2-ol.
4. Whatman 3MM paper.

2.2. PCR Sequencing Reagents

1. 6X Reaction buffer: 240 mM Tris-HCl, pH 7.5, 150 mM MgCl$_2$, 300 mM NaCl, 60% DMSO.
2. Chase mix: 0.25 mM dNTPs, 50 mM NaCl, 10% DMSO.
3. SSB 0.5 µg/µL (USB, Cambridge, UK, #70032).
4. 100 mM Dithiothreitol (DTT). Oligonucleotide sequencing primer (one of the PCR primers or nested internal primer) 0.1 µg/µL in dH$_2$O.
5. ^{35}S dATP-αS (SJ 1304 Amersham International, Amersham, Bucks, UK).
6. ddG Termination mix: 80 µM dGTP, 80 µM dATP, 80 µM dCTP, 80 µM dTTP, 8 µM ddGTP, 50 mM NaCl, 10% DMSO.
7. ddA Termination mix: 80 µM dGTP, 80 µM dATP, 80 µM dCTP, 80 µM dTTP, 8 µM ddATP, 50 mM NaCl, 10% DMSO.
8. ddT Termination mix: 80 µM dGTP, 80 µM dATP, 80 µM dCTP, 80 µM dTTP, 8 µM ddATP, 50 mM NaCl, 10% DMSO.
9. ddC Termination mix: 80 µM dGTP, 80 µM dATP, 80 µM dCTP, 80 µM dTTP, 8 µM ddCTP, 50 mM NaCl, 10% DMSO.
10. Formamide stop mix: 95% formamide, 20 mM EDTA, 0.05% bromophenol blue, 0.05% xylene cyanol.
11. T7 DNA Polymerase (Sequenase USB, Cambridge, UK).
12. Proteinase K (0.2 µg/µL) Sigma Chemical Co., Poole, UK.

3. Method

3.1. Preparation of PCR Products

1. Perform the amplification reaction, and separate the PCR products on 1% agarose gel (*see* Note 2).
2. Visualize the PCR products under a UV transilluminator, and make a cut in the gel in front of the required band.
3. Place Whatman 3MM paper in the cut, and run the gel at 130 V for 30 s.
4. Remove paper from slit, and place the Whatman paper in a 500-mL Eppendorf tube.
5. Repeat steps 3 and 4 a further six to eight times to remove all required PCR product from the gel. This may be monitored by examining the gel under UV for absence of the PCR product (*see* Note 3).
6. Place all Whatman papers in a single 500 µL Eppendorf tube, and make a hole at the bottom of the tube.
7. Place this tube inside a larger 1-mL Eppendorf tube, and centrifuge for 1 min at 12,000g in a microcentrifuge to collect the eluted PCR product in the electrode buffer, 1X TBE.

8. Add an equal volume of butan-2-ol, and vortex briefly. Allow to settle, and discard the upper organic layer. Repeat until a desired volume is obtained (approx 30–40 μL).
9. Wash the concentrate twice with water-equilibrated diethyl ether, and remove excess traces of ether by evaporation using a rotary evaporator.
10. Concentrate the DNA by freeze-drying and reconstituting in an appropriate volume of water, for example, 12 μL.

3.2. PCR Sequencing Reaction

Before beginning sequencing reaction, thaw ddNTP termination mixes, radiolabel and DTT.

1. Add the following to a 500 μL Eppendorf tube:
   a. 2 μL oligonucleotide primer (10 μ*M*) (*see* Note 4).
   b. 2.5 μL concentrated PCR fragment (100–200 ng).
   c. 1 μL 6X reaction buffer.
   d. 0.5 μL SSB (USB #70032).
2. Heat the mix to 96°C for 3 min while adding the following:
   a. 1 μL 100 m*M* DTT.
   b. 1 μL ^{35}S dATP-αS (SJ 1304 Amersham International).
   Mix well and freeze immediately by plunging in liquid nitrogen. While this is in the liquid nitrogen, dilute the T7 Polymerase (Sequenase) 1:8 with water.
3. Remove the tube from the liquid nitrogen, and when the mix is nearly thawed, add 2 μL modified T7 Polymerase (Sequenase USB).
4. Mix well and immediately add 2 μL to each of the four termination mixes (A, C, G, T) and mix well.
5. Incubate at room temperature for 5 min followed by 38°C for 5 min.
6. Add 2 μL of chase mix to each tube and incubate at 38°C for 15 min.
7. Add 4 μL formamide stop mix (from Sequenase kit, USB).
8. Add 0.5 μL of Proteinase K (0.2 μg/μL) and incubate at 60°C for 20 min.
9. Microcentrifuge briefly, boil for 4 min, and load sequencing gel immediately (*see* Notes 5 and 6) (*see* Chapter 1).

4. Notes

1. Most of the materials and reagents used in this protocol are available from Sequenase kit version 2 (USB).
2. Much of the success of this and many other direct sequencing protocols lies in the initial purification techniques of the PCR product from either the reaction mix or from the gel used in the separation. In addition, it is essential that a homogeneous PCR product be used in the sequencing reactions. Purification of PCR product is also possible using a variety of methods, many of which are sold commercially. Protocols detailed in Chapter 2 also describe a number of purification methods.

3. The purification of the PCR product may be monitored by examination of the removal of DNA from a particular region in the gel. When no DNA appears left at the gel position, then the next step of the protocol may be undertaken.

4. Using one of the PCR primers as the sequencing primers may give successful results providing the PCR product is purified from excess dNTPs and primers. In general, however, it would appear that the use of an internal oligonucleotide as the sequencing primer may give a more satisfactory sequence.

5. In general, it is possible to read approx 150–200 bp of sequence from this protocol following an overnight autoradiographic exposure, although staggered gel loadings may increase this by 20–40 bp.

6. Sequencing reactions may be frozen and stored at –20°C for approx 1 wk. However, for best results, it is advisable to separate the products by electrophoresis as soon as possible following the sequencing.

References

1. Saiki, R. K., Scharf, S., Faloona, F., Mullis, K. B., Horn, G. T., Ehrlich, H. A., and Arnheim, N. (1985) Enzymatic amplification of β globin sequences and restriction site analysis for diagnosis for sickle cell anaemia. *Science* **230,** 1350–1354.

2. Erlich, H. A., Gelfrand, D., and Sninsky, J. J. (1991) Recent advances in the polymerase chain reaction. *Science* **252,** 1643–1652.

3. White, B. A. (ed.) (1993) *PCR Protocols: Current Methods and Applications.* Humana, Totowa, NJ.

4. Rapley, R., Flora, P. S., Walsh, D. J., and Walker, M. R. (1993) Nucleotide sequence analysis of CDR3 elements of a panel of anti-peptide monoclonal antibodies recognising parathyroid hormone related protein. *Immunology* **78,** 379–386.

5. Bevan, I. S., Rapley, R., and Walker, M. R. (1992) Sequencing of PCR-amplified DNA. *PCR Methods & Applications* 1 (4): 222–228.

6. Sanger, F., Nicklen, S., and Coulson, A. R. (1977) DNA sequencing with chain terminating inhibitors. *Proc. Natl. Acad. Sci. USA* **74,** 5463–5467.

7. Winship, P. R. (1989) An improved method for direct sequencing PCR amplified material using dimethyl sulphoxide. *Nucleic Acids Res.* **17,** 1266.

14

PCR Sequencing with the Aid of Detergents

**Harald Petry, Barbara Bachmann,
Wolfgang Lüke, and Gerhard Hunsmann**

1. Introduction

The polymerase chain reaction (PCR) allows the rapid detection of infectious viruses or other microorganisms as well as the cause of genetic defects. By performing sequence analysis afterward, important additional information on the PCR products is obtained. Often direct sequencing of the PCR products is preferred over the more time-consuming and expensive procedure of cloning prior to sequencing.

The current sequencing protocol is based on the original Sanger et al. method *(1)*. This technique was first described for single-stranded DNA, and then adapted for sequencing short double-stranded PCR products. DNA fragments of about 200–300 bp may usually be efficiently amplified by PCR. A major difficulty of sequencing short double-stranded PCR products is the strong tendency of the template DNA to reanneal. It has been proven that this difficulty can partially be overcome by adding dimethyl sulfoxide (DMSO) to the sequencing reaction *(2)*. However, despite the addition of DMSO, we have sometimes observed overlaying sequences *(3)*. Therefore, we replaced DMSO by one of the two anionic detergents NP 40 or Tween 20, the diminishing effects of which on the formation of DNA loops are described by Innis et al. *(4)*. Furthermore, there are indications that anionic detergents stabilize polymerases and, therefore, increase the polymerase activities *(5)*.

In the current protocol, we describe a rapid method for the direct sequencing of PCR products. The double-stranded DNA is denatured in the presence of a primer by boiling. Subsequently, the mixture is snap-cooled on dry ice/methanol to prevent reannealing of the template. Instead of DMSO, we used the nonionic detergents NP 40 or Tween 20 to support the prevention of reannealing

From: *Methods in Molecular Biology, Vol. 65: PCR Sequencing Protocols*
Edited by: R. Rapley Humana Press Inc., Totowa, NJ

and loop formation after denaturation of the template DNA. We have found that direct sequencing without DMSO or any detergent frequently leads to unreadable gels *(2)*. More satisfactory results were obtained when the sequencing reactions were carried out in the presence of 10% DMSO. However, most readable sequences were seen when the reactions were performed in the presence of 0.5% Tween 20 or 0.5% NP 40, or when both detergents were employed. NP 40 alone or in combination with Tween 20 strongly enhances the intensity of the resulting bands on an X-ray film. Furthermore, the frequency of unspecific sequencing products was decreased by employing NP 40 alone or in combination with Tween 20.

2. Materials

All solutions must be prepared from molecular-biology-grade reagents and sterile distilled water. Most of the reagents used for the direct sequencing of PCR amplification products are components of commercially available kits. All solutions must be kept frozen at −20°C if not mentioned otherwise in the text.

2.1. Purification of the PCR Products

1. 2% Agarose gel in TBE (0.1 μM Tris base, 0.1 μM boric acid, 0.1 mM EDTA, pH 8.0), stained with ethidium bromide.
2. Commercially available DNA extraction kit (Quiaex, Diagen, Düsseldorf, FRG; Geneclean, Bio 101 Inc., La Jolla, CA).

2.2. Sequencing of the PCR Products

1. For each reaction, 5 pmol of the respective sequencing primer, synthesized on a PCR-Mate EP™ synthesizer (Applied Biosystems, Weitersfadt, Germany). The primers have to be purified by chromatography on NAP 10 columns (Pharmacia, Freiburg, FRG). The primers should be stored in small aliquots at −20°C. Repeated freezing and thawing will decrease the priming efficacy in the sequencing reaction.
2. Detergents: 0.5% Tween 20 or 0.5% NP 40. Prepare detergent solutions early. Commercially available solutions are most highly concentrated and difficult to dissolve in H_2O.
3. Dry ice in methanol.
4. Annealing buffer: 40 mM Tris-HCl, pH 7.5, 25 mM $MgCl_2$, 50 mM NaCl.
5. For each reaction, 10 μCi ^{35}S-dATP, 1200 Ci/mmol (NEN, DuPont, Bad Homburg, FRG). Precaution is necessary when handling the radioisotope ^{35}S-dATP. It is also useful to aliquot the labeled nucleotide in small portions because repeated freezing and thawing decrease the incorporation. Mind the half-life time of ^{35}S: the older the isotope, the more ^{35}S-dATP has to be employed in the sequencing reaction, which may lead to poor results.
6. Pyrophosphatase: 5 U/mL in 10 mM Tris-HCl, pH 7.5, 0.1 mM EDTA.

7. Sequenase version 2.0 T7 DNA polymerase, 13 U/μL (US Biochemicals/Amersham, Braunschweig, FRG), in 20 mM KPO$_4$, pH 7.4, 1 mM DTT, 0.1 mM EDTA, 50% glycerol. We recommend storing the Sequenase in small aliquots at –20°C for a period of up to 3 mo or at –80°C for a longer time to sustain full stability of the polymerase.
8. 0.35 μL Sequenase diluted in 1.95 μL 10 mM Tris-HCl, pH 7.5, 5 mM DTT, 0.5 mg/mL BSA.
9. Sequenase mix: 1.95 μL Sequenase dilution buffer, 0.15 μL Pyrophosphatase, 0.35 μL Sequenase.
10. Reaction mix: 1 μL 0.1M DTT, 1 μL ^{35}S-dATPαs, 2 μL Sequenase mix.
11. The dideoxy mixtures contained either 0.5% Tween 20 or 0.5% NP 40. Both detergents together have a concentration of 0.5%.

	A mix	C mix	G mix	T mix
0.1M dCTP	0.8 μL	0.8 μL	0.8 μL	0.8 μL
0.1M dGTP	0.8 μL	0.8 μL	0.8 μL	0.8 μL
0.1M dTTP	0.8 μL	0.8 μL	0.8 μL	0.8 μL
10 mM ddATP	0.8 μL	—	—	—
10 mM ddCTP	—	0.8 μL	—	—
10 mM ddGTP	—	—	0.8 μL	—
10 mM ddTTP	—	—	—	0.8 μL
1.0M NaCl	50.0 μL	50.0 μL	50.0 μL	50.0 μL
H$_2$O	946.8 μL	946.8 μL	946.8 μL	946.8 μL

12. Chasing mix: 50.0 μL 1.0M NaCl, 2.5 μL 0.1M dATP, 2.5 μL 0.1M dCTP, 2.5 μL 0.1M dGTP, 2.5 μL 0.1M dTTP, 940.0 μL H$_2$O.
13. Stop mix: 95% formamide, 20 mM EDTA, 0.05% bromophenol blue, 0.05% xylene cyanol FF. N.B.: Formamide is toxic.

2.3. Separation of the Sequencing Products

Sequencing products are separated on a polyacrylamide gel (6%) containing 8M urea, 0.2–0.4 mm thick. N.B.: Polyacrylamide is highly toxic. Also take care when you treat the sequencing glass plates with toxic Repel-Silane.

3. Methods
3.1. Separation of the PCR Products

1. Separate the PCR products on a 2% agarose gel, and stain it with ethidium bromide.
2. Estimate the amount of DNA from the fluorescence intensity of the visible bands. This step is highly important, since the amount of DNA employed strongly influences the success of the direct sequencing.
3. Cut out the visible bands, and extract the DNA with a commercially available kit. We have obtained the best results with DNA extraction kits containing glass-milk as DNA adsorbens. However, the solubilization of the DNA from the glass-milk requires special attention, since an incomplete separation may influence the sequencing reaction.

3.2. Direct Sequencing of PCR Products

1. For each reaction, apply the range of 10–50 ng DNA, corresponding to about 0.25 pmol. It has to be mentioned that the DNA concentration for the current method is very important, and the range of tolerance is small.
2. Add 5 pmol of the respective sequencing primer, and dry the mixture at 95°C. The incubation time is dependent on the volume in which the DNA is present. We solubilized the DNA in 2–20 µL H_2O dependent on the PCR efficacy. The procedure then takes 10–30 min.
3. Add 6 µL annealing buffer including the respective detergent with a final concentration of 0.5%. Incubate this mixture at 95°C for 3 min, and immediately chill it on dry ice/methanol. The very rapid cooling is essential to prevent reassociation of the template DNA that can inhibit the sequencing reaction.
4. In the meantime, prepare four reaction tubes. Each reaction tube should contain 2 µL of one of the four dideoxynucleotides mixtures: ddATP, ddCTP, ddGTP, or ddTTP. It is recommended to prepare these four tubes at the beginning of the procedure and store them on ice until required.
5. Add 4 µL of the reaction mixture to the DNA/primer mixtures. Concentrate the total volume at the bottom of the tube by spinning down the reaction mixture at high speed for a few seconds in a bench-top centrifuge.
6. Add 2.3 µL of this solution to each of the four prewarmed dideoxynucleotide mixtures, and incubate them at 37°C for 5 min.
7. Immediately add 2 µL chasing mix to each reaction, and incubate it at 37°C for a further 5 min.
8. Finally, stop the reactions by adding 4.5 µL stop mix. At this stage, the mixtures can be stored at –20°C for a week.

3.3. Separation of the Sequencing Products

1. Heat the sequencing mixture at 95°C for 3 min before running the gel.
2. Carefully apply 4.5 µL of each sequencing reaction to the 6% polyacrylamide gel containing 8*M* urea.
3. Stop the gel until the bromophenol blue reaches the end of the gel, which takes about 3 h at 2000 V. Most sequences longer than 150 bp have to be resolved on gels that run at least 5 h constantly at 2000 V.
4. Transfer the gel on Whatmann 3MM paper (Whatmann, Clifton, NJ). Cover the gel with a plastic film, and dry it.
5. Autoradiograph the gel at room temperature for 12–72 h. The time of exposure is dependent on the efficacy of the sequencing reaction.

4. Notes

1. The size of template should be in the range of 100–800 bp. The efficacy of the protocol decreases with shorter or longer templates.
2. The primer should not contain sequences that could bind to repeated sequences within the template DNA.

3. Signals in all four lanes ("A," "C," "G," "T") from the bottom to the top of the gel can often be explained by unsuitable primer sequences. Primers working well in the PCR are not necessarily suitable sequencing primers.

4. Faint signals on the X-ray film can often be explained by an old charge of ^{35}S-dATP. The older the labeled nucleotide, the higher the concentration of unlabeled dATP in the sequencing reaction.

5. The intensity of the bands is also dependent on storage conditions of the buffers, mainly the Sequenase and the nucleotides. Repeatedly thawing and freezing will lead to hydrolysis of the nucleotides, which causes a decreased incorporation rate.

6. No signals on the X-ray film can often be explained by too high an amount of DNA in the reaction. The current protocol is very useful to sequence 10–50 ng DNA.

7. Overlaying sequences can be explained by an insufficient concentration of detergent in the sequencing mix. This can be traced back to insufficient solubilization of the detergent.

8. Also, overlaying sequences can often be explained by the genotypic variability present in the PCR products.

References

1. Sanger, F., Nicklen, S., and Coulson, A. R. (1977) DNA sequencing with chain-terminating inhibitors. *Proc. Natl. Acad. Sci. USA* **74**, 5463–5467.

2. Bachmann, B., Lüke, W., and Hunsmann, G. (1990) Improvement of PCR amplified DNA sequencing with the aid of detergents. *Nucleic Acids Res.* **18(5),** 1309.

3. Winship, P. R. (1989) An improved method for directly sequencing PCR amplified material using dimethyl sulphoxide. *Nucleic Acids Res.* **17**, 1266.

4. Innis, M. A., Myambo, K. B., Gelfand, D. H., and Brow, M.-A. D. (1988) DNA sequencing with Thermus aquaticus DNA polymerase and direct sequencing of polymerase chain reaction amplified DNA. *Proc. Natl. Acad. Sci. USA* **85**, 9436–9440.

5. Wu, A. M. and Cetta, A. (1975) On the stimulation of viral DNA polymerase activity by nonionic detergent. *Biochemistry* **14(4)**, 789–795.

15

Direct Sequencing with Highly Degenerate and Inosine-Containing Primers

Zhiyuan Shen, Jingmei Liu, Robert L. Wells, and Mortimer M. Elkind

1. Introduction

Among the many techniques of cloning new genes, one approach involves degenerate primers *(1–7)*. The approach usually requires three steps:

1. Using degenerate primers to amplify part of the gene of interest by PCR: The degenerate primers' sequences may be designed from known protein sequences or conserved regions of a gene family (e.g., *2,4*). Because deoxyinosine can base pair with all of the four deoxyribonucleotides it has been substituted for specific nucleic acids in degenerate primers to reduce the number of different primer sequences that would otherwise be needed in the reaction *(2,7,8)*.
2. A determination of which amplified PCR product(s) is from the gene of interest: If the target gene and the primers are only partially homologous, a moderate annealing stringency in the PCR reaction is usually necessary to obtain amplification. Moderate stringency may result in multiple PCR products. Although from the size of the PCR products it may be possible to predict which is from the gene of interest, sequencing analysis of the PCR products may be required.
3. The screening of a cDNA library using the correct PCR product as a probe and cloning the gene of interest.

Sequencing the amplified PCR product is one of the most important steps in this approach to gene cloning. To sequence the PCR fragment amplified by degenerate inosine-containing primers, the PCR fragment may be cloned into a sequencing vector, such as M13 bacteriophage. Sequencing is straightforward if primers specific to the vector are used. Theoretically, this method allows any unknown cloned DNA fragment to be sequenced. However, the *Taq* polymerase, which is used to amplify the target fragment, is thought to have relatively high misincorporation rates for dNTPs, $\sim 10^{-4}$. Hence, it is possible that a

From: *Methods in Molecular Biology, Vol. 65: PCR Sequencing Protocols*
Edited by: R. Rapley Humana Press Inc., Totowa, NJ

copy of the product may contain one or more incorrect nucleotides. If such a copy has been cloned into the sequencing vector, the resulting sequencing data would be incorrect for that particular clone. Direct sequencing of PCR products can circumvent this problem because most of the fragments are exact replicas of the target molecule. Thus, the majority of the products used for sequencing would have the right nucleotide at a specified position and result in the correct sequencing ladder. Also, direct sequencing of PCR products avoids the time-consuming cloning of PCR products, and most of the available direct PCR sequencing protocols require relatively small amounts of template.

Many protocols are available for the direct sequencing of PCR products. Most of the protocols require specific sequencing primers. However, this means at least part of the specific base sequence of the template is needed. This requirement may not be met and, in many cases, information about the internal sequence of a gene may be lacking. This shortcoming may apply to the PCR products of degenerate inosine-containing primers of the cDNA of a new gene. Therefore, one may be forced to use the same degenerate inosine-containing primers that were used in the PCR step for direct sequencing. When primers have low degeneracy, they may be treated as sequence-specific primers, and some of the direct-sequencing protocols, such as those described in this volume, may be used with success. When only highly degenerate inosine-containing primers are available, these methods may not succeed.

To sequence a PCR product amplified via the use of a highly degenerate inosine-containing primers, several general factors must be kept in mind.

1. The sequencing primer(s) must anneal specifically to one site on the DNA fragment that is to be sequenced, i.e., a secondary annealing site must be avoided. Therefore, stringent primer annealing temperatures are necessary.
2. A sufficient quantity of the specific primer should anneal to the correct site. Consequently, the primer annealing temperature cannot be too high.
3. Reassociation of the double-stranded DNA template should be minimized. This requirement generally can be met by using optimal PCR protocols for the sequencing reactions.

To carry out the requirements above, a primer-labeling method, in which the primer is labeled at the 5'-end, may be worth considering. Linear PCR is used to generate the labeled dideoxynucleotide-terminated sequences *(9,10)*. The use of this method minimizes problems of template reassociation and/or mismatching of the primer, because the annealing time is relatively short. Also, the annealing temperature is higher than it would be in most protocols that use DNA polymerases other than *Taq*, such as T4 DNA polymerase, but the method requires a 5'-end-labeling step for which ^{35}S is generally not suitable compared to ^{32}P because of its lower specific activity and the lesser efficiency with which some enzymes label 5'-ends with α-^{35}S-ATP vs α-^{32}P-ATP. Only ^{32}P can be used, even though its greater radiation hazard owing to its higher β-particle emission and its shorter half-life make it less convenient. Furthermore, when highly degenerate primers are used, higher primer concentrations in the reaction mixture are needed to insure that sufficient specific priming will occur. The preceding increases the hazard as well as the cost.

To assure that sequencing primer(s) anneal to a DNA template specifically, to eliminate the need for 5'-end labeling, and to avoid reassociation of the double-stranded DNA template, a two-step cycle-sequencing protocol is described to sequence products amplified with degenerate inosine-containing primers. This method uses the same degenerate primers that were used in PCR amplification. The method can be broken down into two steps of linear PCR. The first step is for labeling the primers, and the second is for the random dideoxy-termination. As shown in Fig. 1, in the first step, primers were extended and labeled with α-^{35}S-dATP. The extension is limited and performed under conditions of high stringency, low dNTP concentration, and a short interval, so that the specific primer in the mixture is favored and a limited length of primer extension is achieved. In the second step, dideoxynucleotide terminations are effected at a more stringent elevated annealing/elongation temperature. The result is that only the extended and labeled primers enter into the termination reactions.

We have used this method to sequence amplified cytochrome p450 cDNA fragments with a highly degenerate inosine-containing primer *(1,2)*. In our case, a set of degenerate primers was used to amplify a presumably novel cytochrome p450 gene(s). The upstream sense primer was a mix of 192, 20 mer, containing three inosines, which theoretically could anneal to 12,288 different sequences. The downstream, antisense primer was a mix of 144, 23 mer, containing five inosines or 147,456 different possible sequences.

In what follows, we will only describe the sequencing reactions. Procedures for sequencing gel electrophoresis can be found in Chapter 1.

2. Materials

1. A thermal cycler: Cetus Perkin-Elmer Model 480 (*see* Note 1).
2. 0.5-mL PCR tubes.
3. Mineral oil.
4. Gel-purification kits/reagents, such as: QIAEX Gel Extraction Kit (Qiagen #20020, Chatsworth, CA) or QiaQuick Gel Extraction Kit (Qiagen #28704).
5. All buffers and solutions must be free of DNase.
6. α-^{35}S-dATP (10 μCi/μL 1000 Ci/mmol) (Amersham Corp).
7. Sequencing primers (degenerate primers) dissolved in H$_2$O, or 0.1X TE buffer.

The sequencing reaction reagents can be homemade. However, we recommend purchasing them from a commercial company to ensure uniform perfor-

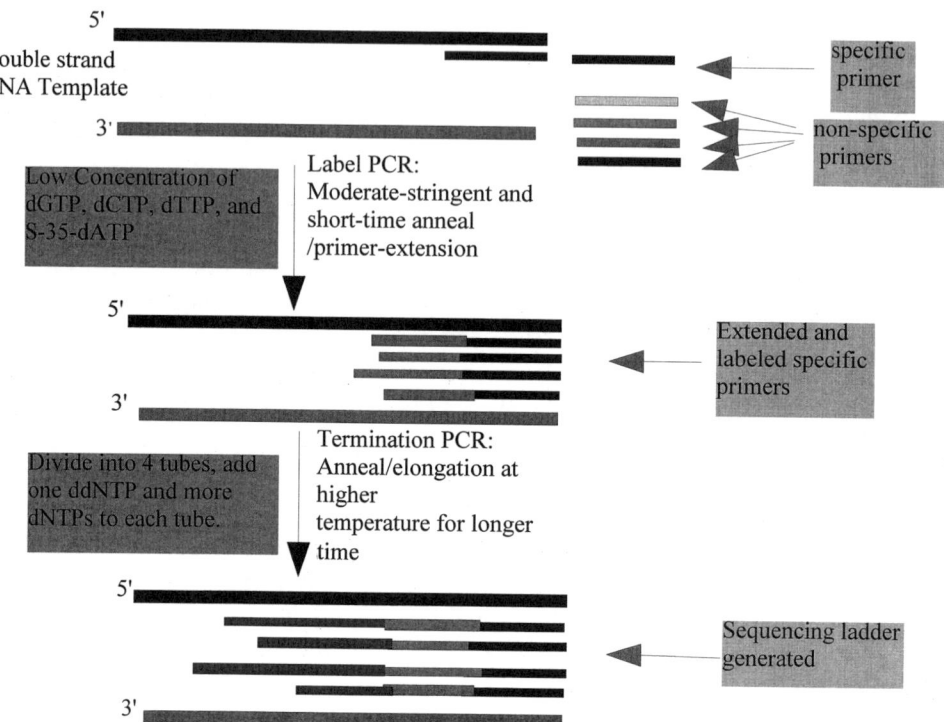

Fig. 1. Procedure of cycle sequencing with degenerate inosine-containing primers. Two linear PCR steps are involved. (1) Label PCR uses low dNTP concentrations, a low temperature, and short times for primer annealing/elongation to produce incomplete extension of specific primers. As a result, specific primers are labeled and extended. The extended and labeled primers have a higher melting temperature than the native printers. (2) Termination PCR using a higher annealing/elongation temperature and is performed with higher dNTP concentrations and in the presence of ddNTPs. Only the extended and labeled primers are involved in the termination reaction.

mance. In the following materials, we include the catalog number for US Biochemicals (Cleveland, OH).

8. Reaction buffer (USB #71030): 260 mM Tris-HCl, pH 9.5, 65 mM MgCl$_2$.
9. ΔTaq DNA polymerase (USB #71059) or Taq DNA polymerase, (USB# 71057): 32 U/μL.
10. Taq DNA polymerase dilution buffer (USB #71051): 10 mM Tris-HCl, pH 8.0, 1 mM 2-mercaptoethanol, 0.5% Tween-20, and 0.5% Nonidet P-40.
11. Four separate primer label mixes:
    a. dGTP label mix: 3.0 μM (USB #71034).
    b. dATP label mix: 3.0 μM (USB #71036).

    c. dTTP label mix: 3.0 μ*M* (USB #71037).

    d. dCTP label mix: 3.0 μ*M* (USB #71038).

12. Four separate termination mix:

    a. ddG terminator mix: 15 μ*M* each dGTP, dATP, dTTP, dCTP, and 22.5 μ*M* ddGTP (USB #71020).

    b. ddA termination mix: 15 μ*M* each dGTP, dATP, dTTP, dCTP, and 300 μ*M* ddATP (USB #71035).

    c. ddT termination mix: 15 μ*M* each dGTP, dATP, dTTP, dCTP, and 450 μ*M* ddTTP (USB #71040).

    d. ddC terminator mix: 15 μ*M* each dGTP, dATP, dTTP, dCTP, and 75 μ*M* ddCTP (USB #71025).

13. Stop/gel-loading solution (USB #70724): 95% formamide, 20 m*M* EDTA, 0.05% bromophenol blue, and 0.05% xylene cyanol FF.

14. 1X TE buffer: 10 m*M* Tris-HCl, pH 8.0, 1 m*M* EDTA.

3. Methods

3.1. Preparation of DNA as a Sequencing Template (see Note 2)

1. After gel electrophoresis, PCR fragment(s) of interest is cut out from the gel.

2. DNA in the gel is purified with the Qiagen gel-purification kit, and final PCR products are resuspended in a proper amount of 0.1X TE buffer.

3. To estimate the amount of PCR product, run an aliquot of the PCR products on an agarose gel. The amount of DNA may be estimated by a comparison with the amount of DNA that was used in the mol-wt ladder.

In the following steps, always keep tubes on "ice," unless otherwise indicated.

4. Prepare the following labeling PCR mix (*see* Note 3):

H$_2$O	0–8 μL
DNA (in 0.1X TE) (need total of 25–100 ng)	1–9 μL
Reaction buffer	2 μL
Degenerate primers (5–200 μ*M*)	1 μL
dGTP label mix	1 μL
dCTP label mix	1 μL
dTTP label mix	1 μL
α-^{35}*S*-dATP (10 μCi/μL >1000 Ci/mmol)	0.5 μL
Taq DNA polymerase (4 U/μL) (diluted in *Taq* dilution buffer)	2 μL
Total volume	17.5 μL

Cover the label PCR mix with 10–20 μL of mineral oil.

3.2. Labeling PCR (see Note 4)

1. Run the following PCR program: presoak at 94°C for 3–5 min followed by 45 cycles of 95°C for 30 s and 52°C for 30 s.

2. Transfer 15–16 μL of the above labeled mixture to a new tube. Avoid carryover of any mineral oil. This can be done easily by putting the pipeting tip directly below the oil without touching the wall of the tube.

3. Optional (*see* Note 5): Load 1–2 μL with 1 μL of gel-loading buffer to a sequencing gel to check the labeling efficiency.
4. Termination PCR mix: For each of the labeled mixes, prepare four tubes labeled as "G," "A," "T," and "C." To each of the tubes, add 4 μL of termination mix "G," "A," "T," or "C" (this can be done toward the end of label-PCR procedure). Add 3.5 μL of the label mix to each of the tubes. Cover the termination PCR mix with 8–10 μL of mineral oil.
5. Termination PCR: Cycle between 95°C for 30 s and 72°C for 90 s (*see* Note 6).
6. While the termination PCR is under way, prepare four clean 0.5-mL tubes labeled "G," "A," "T," or "C." To each of them add 4 μL stop/gel-loading solution.
7. Transfer 6–7 μL of termination mix to these tubes with the stop/loading solution. Avoid carryover of mineral oil. Mix and spin down briefly. Store at –20°C (good for up to 1 mo). These samples are ready for the sequencing gel (use 3 μL to load a gel). *See* Chapter 1.
8. Sequencing results: run sequencing gel, perform autoradiography, and read the sequence (*see* Note 7).

4. Notes

1. Cetus Perkin-Elmer thermal cycler Model 9600 also may be used. If it is, use 0.1-mL tubes; no mineral oil on the top of the reaction solution is needed. The PCR program should be adjusted accordingly in the procedure.
2. Other methods of DNA preparation are also acceptable as long as "clean" DNA is obtained.
3. Other α-^{35}S-labeled nucleotides may also be used, but the label mix must be changed accordingly. The concentration of the stock of degenerate primers in the reaction is dependent on the degree of degeneracy. In our case, the stock concentration of our >100× degenerate primer was 200 μ*M*. Because radioactive ^{35}S is used for these experiments, always be careful and follow the safety operation procedure for your institute. Check with your radiation safety officer for the authorized amount of radioactivity that you can handle at any one time.
4. Depending on the sequencing primer, the annealing/elongation temperature or time may have to be optimized to give proper primer extension and labeling. The purposes of the label PCR is to have a sufficient amount of specific primer in the primer mix to anneal to a specific site on the DNA template, and to extend the annealed primer for a limited nucleotide length with *Taq* DNA polymerase. The first purpose can be achieved by choice of an optimal annealing temperature and/or time. In our case (200 pmol of the 20 mer with inosine and a degeneracy of more than 100×), we used 52°C, and 30 s. Depending on circumstances, this temperature and the annealing time may need to be adjusted. The method of generating a limited elongated primer plus labeling of the primer is accomplished by using a shorter annealing/elongation time at a suboptimal temperature (for *Taq* activity), but still a stringent temperature for annealing and a low dNTP concentration. In this way, it is not necessary to know the sequence of the downstream flanking region of the sequencing primer.

5. Loading 1–2 μL of labeled mix to run a gel to check that the length of primer extension and label efficiency is optional. This can be run along with the sequencing sample after all the reactions are finished. Using our p450 degenerate sequencing primers under these labeling PCR conditions, we obtained an average primer extension of 15–25 bp.

6. The temperature used for both the annealing of labeled/extended primers to the DNA template for the elongation/termination reaction was 72°C. Only the prelabeled and pre-extended primers, which are the specific primers in the primer mix, would be allowed during the termination/elongation because of the elevated temperature. If more template DNA is available, fewer cycles may be used.

7. Depending on the length of labeled primers, the readable sequence will vary. For our case of highly degenerate inosine-containing primers of p450 genes (*see* Section 1. for a description of our p450 primers, a ladder from 25 bp downstream of the primer was readable up to 300 bp.

8. A similar protocol of this method would be to omit one of the four dNTPs in the label step, and use at least one α-^{35}S-labeled dNTP in the labeling mix. This will give an incomplete elongation of the sequencing primer during the labeling step, because the primer extension will stop at the proper position when the omitted nucleotide is not present. The elongated primers may be labeled if the labeled nucleotide is by chance present between the sequence primer and the omitted nucleotide. This method is useful to sequence DNA when some sequence information immediately downstream from the sequencing primer is available. In such a case, one can decide which nucleotide to omit or to label in the label mix.

References

1. Shen, Z., Liu, J., Wells, R. L., and Elkind, M. M. (1993) Cycle sequencing using degenerate primers containing inosines, *BioTechniques* **15(1)**, 82–89.
2. Shen, Z., Wells, R. L., Liu, J., and Elkind, M. M. (1993) Identification of a cytochrome p450 gene by reverse transcription-PCR using degenerate primes containing inosines. *Proc. Natl. Acad. Sci. USA* **90**, 11,483–11,487.
3. Shen, Z., Liu, J., Wells, R. L., and Elkind, M. M. (1994) cDNA cloning, sequence analysis, and induction by aryl hydrocarbons of a murine cytochrome p450 gene, Cyplbl. *DNA and Cell Biol.* **13(7)**, 763–769.
4. Shen, Z., Denison, K., Lobb, R., Gatewood, J., and Chen, D. J. (1995) The human and mouse homologs of yeast RAD52 genes: cDNA cloning, sequence analysis, assignment to human chromosome 12pl2.2-pl3, and mRNA expression in mouse tissues. *Genomics* **25**, 199–206.
5. Compton, T. (1990) Degenerate primers for DNA amplification, in *PCR Protocol, a Guide to Methods and Applications* (Innis, M. A., Gelfand, D. H., Sninsky, J. J., and White, T. J., eds.), Academic, San Diego, CA, pp. 39–45.
6. Lee, C. C., and Caskey, C. T. (1990) cDNA cloning using degenerate primers, in *PCR Protocol, a Guide to Methods and Applications* (Innis, M. A., Gelfand, D. H., Sninsky, J. J., and White, T. J., eds.), Academic, San Diego, CA, pp. 46–59.

7. Knoth, K. S., Roberds, S., Poteet, C., and Tamkun, M. (1988) Highly degenerate inosine-containing primers specifically amplify rare cDNA using the polymerase chain reaction. *Nucleic Acids Res.* **16,** 10,932.

8. Erlich, H. A., Gelfand, D., and Sninsky, J. J. (1991) Recent advances in the polymerase chain reaction. *Sciences* **252,** 1643–1651.

9. Murray, V. (1989) Improved double strand DNA sequencing using the linear polymerase chain reaction. *Nucleic Acids Res.* **17,** 8889.

10. Smith, D. P., Jonstone, E. M., Little, S. P., and Hsiung, H. M. (1990) Direct DNA sequencing of cDNA inserts from plaques using the linear polymerase chain reaction. *BioTechniques* **9,** 48–52.

16

Determination of Unknown
Genomic Sequences Without Cloning

Jean-Pierre Quivy and Peter B. Becker

1. Introduction

The inherent problems of sensitivity and specificity that one encounters when trying to determine a particular nucleotide sequence directly in its genomic context can be overcome by selective amplification of the region of interest. This amplification of the target DNA is usually achieved by one of two strategies: The relevant piece of DNA may be cloned and therefore amplified in a bacterial cell or, alternatively, the desired fragment may be amplified in vitro using PCR technology. Both strategies have drawbacks. The cloning of a specific genomic sequence is labor-intensive, lengthy, and sometimes even difficult to achieve. The PCR amplification requires that enough sequence information is known to be able to design the two specific amplification primers and is therefore limited to sequencing alleles of already known DNA. There are, however, many cases that would benefit from the determination of unknown genomic sequence close to a known piece of DNA. With a particular cDNA in hand, one may wish, for example, to determine genomic gene sequences, such as the promoter of the gene, its introns, or 5'- and 3'-nontranscribed regions. The protocol presented here uses ligation-mediated PCR (LM-PCR) to amplify unknown genomic DNA next to a short stretch (about 100 bp) of known sequence and details a convenient procedure to determine the new sequence by dideoxy sequencing *(1)*. The procedure may form the basis for "walking sequencing" strategies in order to determine large regions of continuous sequence information starting from a limited piece of known DNA.

The central feature of the LM-PCR technique is the ligation of a known short oligonucleotide, the "linker," to selected ends of genomic DNA fragments (Fig. 1). These generic linker sequences provide the second primer for

From: *Methods in Molecular Biology, Vol. 65: PCR Sequencing Protocols*
Edited by: R. Rapley Humana Press Inc., Totowa, NJ

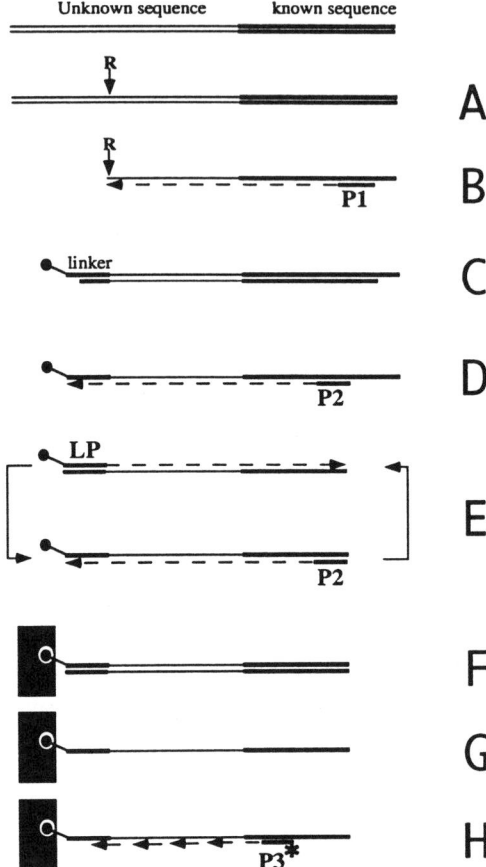

Fig. 1. The use of Linker Tag Selection LM-PCR for genomic dideoxy sequencing. Bold lines denote known sequences, and thin lines the unknown sequences to be determined. The arrows and dotted lines indicate primer extension reactions. R: Strategic restriction site. P1, P2, P3, and LP stand for the primers 1, 2, 3, and the linker primer, respectively. The biotin moiety on the linker is represented by a filled circle, and the streptavidin-coated paramagnetic beads by the shaded boxes. (H): Radiolabeled P3 is used to prime dideoxy sequencing reactions.

amplification of linked fragments in combination with an oligonucleotide based on known sequences. LM-PCR was first introduced for genomic footprinting and chemical sequencing *(2,3)*. The disadvantages of chemical sequencing over the chain termination method *(1)* prompted us to adapt LM-PCR technology for direct dideoxy sequencing of genomic DNA *(4)*. The underlying procedure is derived from a variation of the original LM-PCR protocol called "Linker Tag Selection LM-PCR" *(5)*.

The steps of the reaction are outlined in Fig. 1. A restriction enzyme is selected that cleaves the genomic DNA somewhere within the unknown sequence, but within 1 kb from the known sequence for which the specific primers have been designed. Cleavage creates a defined end (A). The cleaved genomic DNA is denatured, and the gene-specific primer 1 is annealed to the known sequence and extended by a polymerase until the end of the restriction fragment (B). This creates a blunt end to which a short double-stranded linker is ligated (C). The linker DNA consists of two complementary oligonucleotides, the longer one being biotinylated at its 5'-end. After denaturation, the specific primer 2, representing sequences more 3' from primer 1 on the lower strand of the known sequence, is then annealed to the upper strand, which now carries the biotinylated linker oligonucleotide at its 5'-end. The primer is again extended to the end (D) creating a double-stranded fragment that contains a region of unknown DNA flanked by known sequences. The genomic sequences can now be amplified by PCR using a combination of the biotinylated linker primer and primer 2 (E). The biotinylated amplification products are then immobilized on streptavidin-coated paramagnetic beads and purified from the PCR reaction in a magnetic field (F). This step efficiently removes unincorporated primers, which interfere with the subsequent sequencing reaction. The immobilization also facilitates the handling of the template fragments during the subsequent steps and specifically allows the sequencing of a single-stranded template. The immobilized fragments are denatured, and the complementary strand is removed by washes (G). Radioactively labeled specific primer 3, again located 3' to primer 2 on the lower strand, now serves to prime a standard chain termination sequencing reaction that finally generates the sequencing ladder (H). Sequences start to be readable about 25 bases 3' from primer 3, and may extend for up to 1 kb depending on the location of the restriction site and the efficiency of the overall process (Fig. 2A).

Important features of the procedure are the use of a proofreading polymerase for the PCR amplification (the Vent DNA polymerase possesses a 3'–5' exonuclease activity) that decreases the error rate during the amplification, and the use of an enzyme without the 3'–5' exonuclease activity (e.g., Vent Exo⁻) for efficient single primer extensions. The immobilization of the amplified fragments on paramagnetic beads exploiting the strong streptavidin/biotin interaction is crucial for the efficiency of the sequencing reaction, since it allows the efficient removal of interfering primers from the PCR reaction and permits the use of a single-strand template *(4,5)*. The attachment of the sequenced DNA fragment to the solid support does not create a steric hindrance for the polymerase. Frequently, the sequence of the linker oligonucleotide itself can be determined at the very end of the genomic sequence (*see* Fig. 2B).

The length of sequence determined critically depends on the ability to resolve long fragments with single nucleotide resolution, provided that the

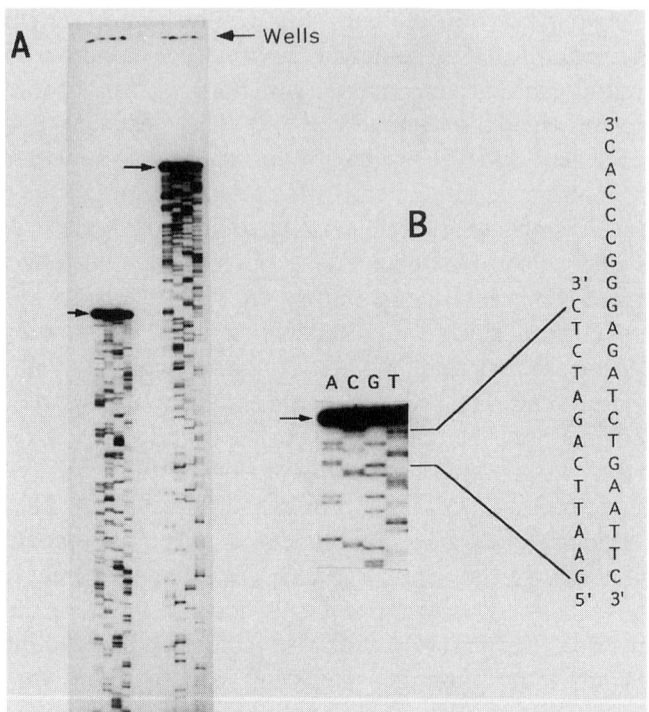

Fig. 2. Genomic sequences of the *Drosophila* hsp27 promoter region obtained following the outlined procedure. Arrows indicate the full-length restriction fragment. **(A)** Long and short gel runs of the same sequencing reaction obtained using the enzyme *Nru*I, which cut 821 bases upstream of the 5'-end of primer 1. **(B)** Upper part of a sequence obtained using the enzyme *Pst*I cutting 332 bases away from the 5'-end of primer 1. The sequence of the linker oligonucleotide at the end of the genomic sequence can be unambiguously identified.

restriction site used for linker ligation is not too close to the known sequence. Fragments of over 800 bp have yielded reliable sequence information (Fig. 2A). Longer sequences can be obtained in walking strategies where the newly determined sequence is in turn used to design further reaching sets of primers. The presented strategy critically relies on the prior identification of a suitable restriction site, ideally between 0.5 and 1 kb away from the known sequence. Too short fragments will yield little new sequence information, whereas large fragments (exceeding 1 kb) do not work, presumably because of decreasing efficiencies in DNA denaturation and primer extensions reactions. Since any kind of restriction enzyme will work, a site can be conveniently identified on a Southern blot testing a small selection of enzymes that cut the genome at a reasonable frequency. Alternatively, a selection of enzymes can simply be tried at random in an LM-PCR sequencing reaction. To increase the chances that the reaction will work, the genomic DNA may be cleaved with a whole cocktail of enzymes that collectively have a high likelihood to produce a suitable restriction fragment.

2. Materials
2.1. Purification and Restriction of Genomic DNA

1. Suspension of nuclei from desired organism (*see* Note 1).
2. 0.5*M* EDTA, pH 8.0.
3. RNase A, DNase free, 10 mg/mL (Boehringer, Mannheim).
4. Aqueous solution of *N*-lauroylsarkosine (sarkosyl), 20% (P/V) (Sigma).
5. Proteinase K, 10 mg/mL (Merck).
6. Phenol, highest quality, neutralized, and equilibrated with TE (10 m*M* Tris-HCl, pH 7.5, 1 m*M* EDTA) (Aurresco).
7. Phenol/chloroform/isoamyl alcohol mixture (25:24:1) (Aurresco).
8. Chloroform/isoamyl alcohol (24:1) (Merck).
9. 0.3*M* Sodium acetate, pH 5.2.
10. Ethanol 100%.
11. Ethanol 80%.
12. TE: 10 m*M* Tris-HCl, pH 7.5, 1 m*M* EDTA.
13. Restriction enzyme with suitable 10X reaction buffer (*see* Note 2).

2.2. Primer Design, Primer Purification, and Annealing of the Linker Primer (see Note 3)

The primers were synthesized on an ABI 394 DNA synthesizer and gel purified (*see* Section 3.2.).

1. Primer 1 (P1): a 18–22 mer with a calculated T_m around 45–50°C. Working concentration: 0.5 pmol/μL.
2. Primer 2 (P2) should have the same melting temperature as the long-linker primer (*see* Note 4) used for the PCR amplification, in our case a 25–27 mer with a calculated T_m of 60–65°C. The GC content is usually between 45 and 55%. It does not need to overlap with P1, but a 5-bp overlap has worked. It should be internal to primer 2 to increase the specificity of the overall reaction. Working concentration: 10 pmol/μL.
3. Primer 3 (P3): a 18–22 mer with a calculated T_m around 45–50°C, but longer oligos with higher GC contents will also work. It should be internal to P2 to increase the specificity. It will be ^{32}P kinased with an SA of 10^7 cpm/pmol (*see* Section 3.3.). Working concentration: 0.2 pmol/μL.

4. Long-linker oligonucleotide: 5' CACCCGGGAGATCTGAATTC 3' (*see* Note 4). It is biotinylated at its 5'-end during synthesis by incorporation of a biotin-2-*o*-propylphosphoramidite. It should be unphosphorylated.
5. Short-linker oligonucleotide: 5' GAATTCAGATC 3', dephosphorylated.
6. Oligo loading mix: 10% glycerol in formamide.
7. Denaturing polyacrylamide gel: 14.5% acrylamide, 0.5% bis-acrylamide, 7M urea, 1X TBE. Size: 25 × 25 × 0.1 cm.
8. Formamide loading buffer: 96% formamide, 0.05% xylene cyanol, 0.05% bromophenol blue, 10 mM EDTA.
9. PE buffer: 50 mM Tris-HCl, pH 7.5, 100 mM NaCl, 1 mM EDTA, 1% phenol (v/v).
10. Chloroform/isoamyl alcohol (24:1) (Merck).
11. 5M LiCl.
12. 1M MgCl$_2$.
13. Ethanol 100%.
14. Ethanol 80%.
15. TBE: 90 mM Tris-borate, 1 mM EDTA.

2.3. Kinasing of Primer 3

1. P3 at 10 pmol/µL.
2. Polynucleotide kinase buffer 10X: 700 mM Tris-HCl, pH 7.6, 100 mM MgCl$_2$, 50 mM DTT.
3. γ^{32}PATP, 5000 Ci/mmol (Redivue, Amersham).
4. Polynucleotide kinase 10 U/µL (NE Biolabs).
5. 50 mM EDTA, pH 8.0.
6. G-25 fine spin columns (Boehringer, Mannheim).

2.4. First Primer Extension

1. P1 at 0.5 pmol/µL in TE, pH 7.5.
2. 1N NaOH.
3. TES buffer: 560 mM TES, free acid (Sigma), 240 mM HCl, 100 mM MgCl$_2$.
4. Vent buffer (10X): 100 mM KCl, 100 mM (NH$_4$)$_2$SO$_4$, 200 mM Tris-HCl, pH 8.8, 20 mM MgSO$_4$, 0.1% Triton X-100 (NE Biolabs).
5. dNTP solution: 10 mM dNTPs, (Boehringer, Mannheim); keep in small aliquots at −20°C. Do not freeze/thaw more than three times.
6. Vent Exo⁻ DNA polymerase, 2 U/µL (NE Biolabs).

2.5. Ligation

1. Ligase buffer (10X): 500 mM Tris-HCl, pH 7.5, 100 mM MgCl$_2$, 100 mM DTT, 10 mM ATP, 250 µg/mL BSA (NE Biolabs).
2. Solution of 40% PEG 8000 (Sigma), filtered through a 0.22-µm filter. It will take some time to dissolve the PEG in water. Incubate at room temperature on a rotating wheel for several hours. It also takes some force to filter the solution using a syringe.
3. T4 DNA ligase, 400 U/µL (NE Biolabs).
4. Annealed linker (*see* Section 3.4.).

2.6. PCR

1. TE: 10 mM Tris-HCl, pH 8.5, 1 mM EDTA, pH 8.5.
2. Phenol/chloroform/isoamyl alcohol (25:24:1) (Aurresco).
3. Solution of 7.5M ammonium acetate containing 25 µg/mL yeast tRNA (Boehringer Mannheim). Crude yeast tRNA has to be cleaned by multiple organic extractions and ethanol precipitation.
4. Vent buffer (*see* Section 2.4.).
5. dNTP solution (*see* Section 2.4.).
6. 100 mM MgSO$_4$.
7. P2 solution, 10 pmol/µL (*see* Section 2.2.).
8. Long-linker primer, 10 pmol/µL (*see* Section 2.2.).
9. Vent DNA polymerase 2 U/µL (NE Biolabs).
10. Perkin-Elmer thermal cycler.
11. PCR tubes (Perkin Elmer).
12. Mineral oil (PCR-grade).

2.7. Tag Selection of the PCR Products

1. Dynabeads M-280 streptavidin (Dynal, Oslo, 10 mg/mL).
2. Magnetic particle concentrator (MPC, Dynal).
3. Phosphate-buffered saline (PBS), pH 7.4.
4. PBS, pH 7.4, 0.01% BSA (molecular-biology-grade).
5. BW solution: a 1:1 mixture of TE and 5M NaCl.

2.8. Sequencing Reaction (see Note 7)

1. Magnetic particle concentrator.
2. 150 mM NaOH, freshly prepared.
3. TE, pH 7.5.
4. Vent buffer (*see* Section 2.4.).
5. Termination mixes made up in 1X vent buffer:
   A-mix: 900 µM ddATP, 30 µM dATP, 100 µM dCTP, 100 µM dGTP, 100 µM dTTP.
   C-mix: 480 µM ddCTP, 30 µM dATP, 37 µM dCTP, 100 µM dGTP, 100 µM dTTP.
   G-mix: 400 µM ddGTP, 30 µM dATP, 100 µM dCTP, 37 µM dGTP, 100 µM dTTP.
   T-mix: 720 µM ddTTP, 30 µM dATP, 100 µM dCTP, 100 µM dGTP, 33 µM dTTP.
6. Labeled P3 (*see* Section 3.3.).
7. Circumvent sequencing buffer 10X: 100 mM KCl, 100 mM (NH$_4$)$_2$SO$_4$, 200 mM Tris-HCl, pH 8.8, 50 mM MgSO$_4$ (NE Biolabs).
8. Triton X-100 solution, 3% (v/v) in water.
9. Vent Exo$^-$ DNA polymerase, 2 U/µL (NE Biolabs).
10. Formamide loading buffer (*see* Section 2.2.).
11. Sequencing gel.
12. Fixing solution: 10% acetic acid, 10% methanol.
13. Dupont NEN reflection™ films and corresponding cassettes.

3. Methods

3.1. Purification of Genomic DNA and Restriction (see Note 1)

1. Isolate nuclei from cells of interest by suitable methods (*see* Note 1), and spin them down to obtain the nuclear pellet.
2. Resuspend the pellet in 1 mL of 0.5*M* EDTA. Avoid harsh vortexing and vigorous pipeting to prevent shearing (*see* Note 1).
3. Add 25 µL RNase A and 25 µL Sarkosyl, mix by inverting the tube, and incubate for 3 h at 37°C on a rotating wheel.
4. Add 25 µL of proteinase K, mix by inverting the tube, and incubate overnight at 37°C on a rotating wheel.
5. Add 1 mL of phenol, and mix by inverting the tube several times. Spin to separate the phases, and collect the lower phase and interphase (the lower phase is the aqueous phase owing to the high density of 0.5*M* EDTA).
6. Repeat step 5, but do not take the interphase.
7. Add 1 mL of phenol/chloroform and collect the upper phase (which is now the aqueous phase).
8. Dialyze overnight (or longer) against TE, pH 7.5, with at least four changes of TE. Avoid a large increase in volume by keeping the dialysis bag tight.
9. Precipitate DNA with 1/10 vol of 0.3*M* Na acetate and 2.5 vol of cold absolute ethanol.
10. Spin at 4°C to collect DNA pellet, wash with 80% ethanol, remove residual ethanol, but do not dry too long since the DNA will be difficult to redissolve.
11. Dissolve DNA in TE, pH 7.5, and store at 4°C.
12. Restriction digest of genomic DNA: Digest 10 µg of DNA with 1 U/µg of restriction enzyme according to the manufacturer's recommendation (*see* Notes 2 and 4) for at least 3 h (up to overnight).
13. Extract DNA once with phenol, once with phenol/chloroform, once with chloroform, and precipitate with 1/10 vol of 0.3*M* Na acetate and 2.5 vol of cold 100% ethanol.
14. Spin at 4°C, wash DNA pellet with 80% ethanol, and remove residual ethanol in the SpeedVac. Again, do not dry too long. Resuspend DNA in TE, pH 7.5, and adjust concentration to 1 µg/µL (OD at A_{260}).
15. Extract once more with chloroform, and remove traces of chloroform in the SpeedVac.

3.2. Purification of Oligonucleotide Primers and Annealing of the Linker Fragment

1. Dry down 75 nmol of the oligonucleotide in the SpeedVac concentrator, and dissolve in 75 µL of oligo loading mix. Heat for 5 min at 75°C, and load 5×15 µL onto a prerun denaturing polyacrylamide gel. In a separate slot load some formamide loading buffer to monitor the run. Electrophorese until the bromophenol marker dye migrated to 2/3 of the gel.
2. By UV shadowing *(6)*, locate the band corresponding to the full-length oligonucleotide, and excise from the gel using a razor blade.
3. Transfer the polyacrylamide gel slice to a 1.5-mL reaction tube containing 1 mL of PE buffer, and incubate overnight at 37°C.

4. Filter the supernatant through a 0.22-μm filter, prewetted with PE, with the help of a 2-mL syringe.
5. Wash another 100 μL of PE buffer through the filter.
6. Extract the pooled PE solutions with chloroform, and dispense 2×450 μL of the upper phase into fresh tubes.
7. Precipitate the oligonucleotides by the addition of 36 μL of 5M LiCl, 4.5 μL of 1M MgCl$_2$, and 1 mL cold 100% ethanol.
8. Mix by vortexing, and let precipitate at $-70°C$ for 15 min.
9. Spin at 4°C for 20 min in a tabletop centrifuge, wash the pellet with 80% ethanol, dry in the SpeedVac, and resuspend the oligonucleotide in TE, pH 7.5.
10. Determine the concentration (OD at A$_{260}$), and dilute some aliquots to the working concentration.
11. Annealing of linker oligos: combine in a 1.5-mL reaction tube: 20 pmol/μL of each linker oligonucleotide in 250 mM Tris, pH 7.5, 5 mM MgCl$_2$. Heat at 95°C for 5 min, transfer to a beaker containing boiling water, and allow to cool slowly at room temperature for 5 h (up to overnight in the cold room). Aliquot the linker solution and store at $-20°C$. Aliquots are only used once and never refrozen.

3.3. Kinasing of Primer 3

1. Combine in a tube: 1 μL of P3 (10 pmol), 3 μL of 10X T4 polynucleotide kinase, and 10.5 μL water.
2. Add 15 μL γ^{32}PATP and 0.5 μL polynucleotide kinase.
3. Incubate for 30 min at 37°C.
4. Add 20 μL of 50 mM EDTA, pH 8.0, and heat at 65°C for 10 min.
5. Purify the labeled oligonucleotide from the unincorporated label by a G-25 spin column.

3.4. First Primer Extension

1. Combine in a tube: 0.5–1 μg of restricted genomic DNA (*see* Section 3.1. and Note 4), 1 μL P1 (0.5 pmol), 1 μL 1N NaOH, and water to 8 μL.
2. Incubate at 65°C for 5 min.
3. Immediately add 2 μL TES buffer and mix.
4. Spin to collect, and incubate for 10 min at room temperature.
5. Add 9 μL of a mix containing 2 μL 10X Vent buffer, 0.4 μL 10 mM dNTPs, 6.6 μL H$_2$O.
6. Incubate at 50°C for 10–20 min.
7. Add 1 μL of the Vent Exo$^-$ (2 U).
8. Incubate for 10 min at 76°C.
9. Chill on ice, spin to collect liquid, and proceed immediately to ligation.

3.5. Ligation

1. Prepare a premix containing 5 μL of 10X ligase buffer, 19 μL of 40% PEG 8000, 5 μL of annealed linker (20 pmol/μL). Mix well by pipeting since the resulting solution is very viscous.

2. Add 29 µL of this premix to the first primer extension reaction (*see* Section 3.4.).
3. Add 1 µL of T4 DNA ligase (400 U), mix well by pipeting, and incubate overnight at 17°C.

3.6. PCR

1. To the ligation reaction, add 150 µL of TE, pH 8.5, and mix by vortexing.
2. Add 150 µL of phenol/chloroform/isoamyl alcohol (25:24:1), and mix by vortexing. Spin 5 min.
3. Collect the upper aqueous phase and transfer to a tube containing 10 µL of 7.5M NH$_4$Ac/yeast tRNA. Add 750 µL of cold 100% ethanol, mix by vortexing, and let precipitate for 15 min at –20°C.
4. Centrifuge for 20 min at 4°C, remove the supernatant, wash the pellet with 500 µL 80% ethanol, and dry in the SpeedVac.
5. Dissolve DNA in 20 µL of water, transfer to a 500-µL PCR tube, and keep on ice.
6. Prepare a premix containing 5 µL 10X Vent buffer, 1 µL 10 mM dNTPs, 1 µL 100 mM MgSO$_4$ (*see* Note 5), 1 µL P2 solution, 1 µL long-linker primer, and 19.5 µL of water.
7. Immediately before use, add 1.5 µL Vent DNA polymerase (3 U) to the premix, and mix by pipeting. Add premix to the PCR tube containing the DNA, add two drops of mineral oil, and keep on ice.
8. Transfer the tube to the thermal cycler (one droplet of mineral oil in sample holders) preheated to 95°C, incubate for 2.5 min at 95°C, and subject to 18 cycles as follows: 95°C for 1 min, 60°C for 2 min (*see* Note 5), 76°C for 3 min. Allow an increase of 5 s/c for the extension step, and end the cycling by a 10-min incubation at 76°C.
9. The PCR reaction can be used immediately for the labeling step or stored frozen at –20°C.

3.7. Linker Tag Selection of the PCR Product

1. Resuspend the Dynabeads M-280 streptavidin well, and take 50 µL (500 µg beads) into a 1.5-mL reaction tube.
2. Concentrate in the magnetic rack (MPC, Dynal), and remove the supernatant (0.01% BSA) (*see* Note 6).
3. Wash the beads with 100 µL of PBS, pH 7.4, concentrate, and remove the supernatant. Repeat this step.
4. Resuspend beads in 100 µL of PBS/0.01% BSA by pipeting up and down until a homogenous suspension is achieved. Concentrate and remove supernatant.
5. Wash the beads with 100 µL BW solution by pipeting up and down until no aggregates are seen. Concentrate again. Repeat this step.
6. Finally resuspend the beads in 100 µL of BW solution. Beads are now ready to be used.
7. Add the PCR reaction to the beads, and mix well. Avoid mineral oil, which spoils the magnetic separation of beads (we do not recommend a chloroform extraction since traces of this solvent adversely affects later steps in the reaction).

8. Incubate the tube on a rotating wheel at room temperature for 30 min. Assure that the beads do not sediment in the tube, but also avoid a spreading of the liquid over the entire tube wall. A suitable agitation can be obtained by adjusting the rotation angle.
9. Concentrate beads, and discard supernatant.
10. Wash the beads with 100 μL of BW solution.

3.8. Template Denaturation and Sequencing Reaction (see Note 7)

1. Resuspend beads well in 100 μL 150 m*M* NaOH by pipeting up and down. Incubate at room temperature for 5 min with occasional gentle agitation, and then for 2 min at 50°C.
2. Spin shortly to collect condensed liquid and beads trapped in the lid. Concentrate beads.
3. Remove the supernatant, and resuspend the beads in 100 μL 150 m*M* NaOH. Spin shortly to collect all the NaOH solution and concentrate beads.
4. Discard supernatant, resuspend the beads in 100 μL TE, pH 8.5, and concentrate again. Repeat this wash with TE.
5. Resuspend the beads in 50 μL of Vent buffer, and keep on ice.
6. Prepare four tubes labeled "A," "C," "G," "T" containing 3 μL of the respective termination mixes (*see* Section 2.8.).
7. Prepare a premix containing 0.2 pmol labeled P3, 1.5 μL 10X circumvent sequencing buffer, 1 μL Triton X-100, and add water to 15 μL.
8. Concentrate beads, and remove supernatant.
9. Resuspend the beads in 15 μL of premix.
10. Transfer 3.5 μL of this bead suspension into each tube containing a termination mix, and mix by pipeting.
11. Heat at 95°C for 10 s, and then transfer to 50°C. Incubate for 10–20 min with occasional gentle resuspension.
12. Add 1 μL of Vent Exo⁻ (2 U) to each tube, mix well, and immediately transfer to 76°C.
13. Incubate for 10 min, and then chill on ice.
14. Spin to collect liquid, concentrate beads, and discard supernatant.
15. Resuspend beads in 50 μL water, spin shortly to collect liquid, and concentrate beads.
16. Carefully remove all liquid, and resuspend the beads in 4 μL of formamide loading buffer/0.15 NaOH (2:1). Mix well to dissolve all aggregates (*see* Note 8).
17. Leave for 5 min at room temperature.
18. Incubate 3 min at 76°C, spin to collect liquid, and chill on ice.
19. Concentrate beads, and transfer supernatant into a fresh tube on ice.
20. Check with a hand monitor that most of the radioactivity is in the supernatant. The bead pellet always contains radioactivity owing to trapped P3. Usually, we do not re-extract the beads.
21. Load on a prerun sequencing gel (*see* Chapter 1) (reactions can be stored at –20°C).
22. Fix gel in fixing solution, dry onto blotting paper, and expose the dried gel to X-ray film in the presence of an intensifying screen. Readable sequences are usually obtained after overnight exposure.

4. Notes

1. The genomic DNA used for genomic sequencing needs to be clean and undegraded. Any shearing of the DNA during preparation and handling before the first primer extension must be avoided. Nicks between the restriction site and the P1 priming site will be converted to blunt ends during the first primer extension and will give rise to a background of dominant bands in all four sequencing lanes. We detail here one particular protocol that consistently yields high-mol-wt genomic DNA of good quality. In principle, other methods can be followed. In any case, we advise to start a DNA prep with a nuclei isolation. Some suggestions for how to prepare nuclei have been described *(7,8)*.

2. Enzymes that produce blunt ends, with 5'- or 3'-overhangs can be used since the fragment is anyway converted to a blunt end after the first primer extension.

3. All oligonucleotides should be gel-purified (*see* Section 3.2.) and stored in water or TE, pH 7.5, at two concentrations: a stock solution (determined after the purification) and a diluted working solution that should not be frozen and thawed more than five times. The sequence-specific primers 1–3 need to be designed for each new sequencing project. The long and short linker oligonucleotides are constant. The distances between primers 1/2 and 2/3 should not be longer than 10 bases; overlaps of up to 5 bp are tolerated, but not required.

4. This protocol was used to sequence in the context of the *Drosophila* genome, which required changes from the original protocol described for mammalian DNA. If sequences are to be determined in the context of the 10 times more complex mammalian genome, a few parameters have to be adjusted: The amount of starting DNA used should be increased about five times, and the longer-linker oligos should be taken from the original procedure *(3)*.

5. The annealing temperature and the optimal magnesium concentration may be optimized specifically for each primer, but the given conditions worked reasonably well for most of the primers we tested.

6. In general, beads are concentrated by placing the 1.5-mL reaction tube on the magnetic rack for 10–20 s. If left for too long, the bead pellets become too tight and therefore difficult to resuspend. The supernatant is removed with a pipet tip with the tube still in the rack. Great care is taken at every step to resuspend the beads well by pipeting the suspension up and down. A drying of the bead pellets should also be avoided.

7. The reagents used for sequencing (*see* Sections 2.8. and 3.8.) are available as a kit (Circumvent sequencing kit, NE Biolabs).

8. Section 3.8., step 19 allows the separation of the labeled fragments from the beads, which facilitates gel loading. However, the presence of the beads in the sample load does not adversely affect the migration of DNA fragments.

References

1. Sanger, F., Nicklen, S., and Coulson, A. R. (1977) DNA sequencing with chain-termination inhibitors. *Proc. Natl. Acad. Sci. USA* **74,** 5463–6467.

2. Pfeifer, G. P., Steigerwald, S. D., Mueller, P. R., Wold, B., and Riggs, A. (1989) Genomic sequencing and methylation analysis by ligation mediated PCR. *Science* **246,** 810–813.
3. Mueller, P. R., Garrity, P. A. and Wold, B. (1993) Ligation mediated PCR for genomic sequencing and footprinting, in *Current Protocols in Molecular Biology* vol. 2 (Ausubel, F. M., Brent, R., Kingston, R. E., Moore, D. D., Seidman, J. G., Smith, J. A., and Struhl, K., eds.), Current Protocols, New York, pp. 15.5.1–15.5.26.
4. Quivy, J.-P. and Becker, P. B. (1994) Direct dideoxy sequencing of genomic DNA by ligation-mediated PCR. *BioTechniques* **16,** 238–241.
5. Quivy, J.-P. and Becker, P. B. (1993) An improved protocol for genomic sequencing and footprinting by ligation-mediated PCR. *Nucleic Acids Res.* **21,** 2779–2781.
6. Sambrook, J., Fritsch, E. F., and Maniatis, T. (1989) *Molecular Cloning—a Laboratory Manual* Cold Spring Harbor Laboratory Press, Cold Spring Harbor, NY.
7. Wu, C. (1989) Analysis of hypersensitive sites in chromatin. *Methods Enzymol.* **170,** 269–289.
8. Bellard, M., Dretzen, G., Giangrande, A., and Ramain, P. (1989) Nuclease digestion of transcriptionally active chromatin. *Methods Enzymol.* **170,** 317–346.

17

DNA Sequencing by the Chemical Method

Eran Pichersky

1. Introduction

The chemical method of sequencing DNA *(1)* has some advantages and some disadvantages compared with the enzymatic method *(2)*. The major disadvantage is that it takes more time to produce the same amount of sequence. This is so for two main reasons. First, the DNA has to be end-labeled and then reisolated prior to the actual chemical sequencing reactions, a process that usually requires an additional day. Also, because more DNA is used in the reaction and because the lower specific activity of the sequenced DNA requires the use of an intensifying screen in the autoradiography, bands are not as sharp as in the enzymatic method and therefore it is difficult to obtain reliable sequence past about nucleotide 250 (unless very long gels are run).

Nevertheless, the chemical method is often useful for several reasons. It enables one to begin sequencing anywhere in the clone where a labelable restriction site occurs without any further subcloning. The sequence thus obtained can then be used to synthesize oligonucleotide primers for enzymatic sequencing. In addition, in cases of regions that give poor results in the enzymatic reactions (because of secondary structures that inhibit the polymerase enzyme), the chemical method almost always resolves the problem and yields the correct sequence.

The chemical sequencing reaction has acquired a reputation of being difficult. We believe this reputation in undeserved. In our hands, the chemical method is consistently successful in producing results as reliable as those obtained by the enzymatic method. It has been our experience that many protocols in molecular biology include unnecessary steps. The likely explanation is probably that when researchers encountered difficulties, they added these steps as a solution to the problem, often on the assumption that the additional steps would not hamper the process, even if they did not help. This is clearly not the case here. In developing

From: *Methods in Molecular Biology, Vol. 65: PCR Sequencing Protocols*
Edited by: R. Rapley Humana Press Inc., Totowa, NJ

the method presented here from pre-existing protocols, we have *eliminated* many steps. In general, we have found that the quality of the sequence has improved with the progressive elimination of these steps. It is still possible that some steps included here are not necessary; certainly no additional steps need to be added. And, of course, the end result has been that the protocol as presented here is very short and the entire process of sequencing (starting with end-labeled DNA) and gel electrophoresis can be accomplished in one (long) day.

2. Materials

1. G Buffer: 50 mM sodium cacodylate, pH. 8.0.
2. CT/C Stop: 0.3M sodium acetate, 1 mM EDTA, pH 7.0.
3. GA Stop: 0.3M sodium acetate, pH 7.0.
4. G Stop: 1.5M sodium acetate, 1M 2-mercaptoethanol, pH 7.0.
5. 10% Formic acid.
6. Dimethyl sulfate (DMS) (*see* Notes 1 and 4).
7. Hydrazine (95% anhydrous) (*see* Notes 1 and 4).
8. 100% ethanol.
9. 5M NaCl.
10. ddH$_2$O (dd = double distilled).
11. Carrier DNA: 1 mg/mL in ddH$_2$O (any DNA will do; we use plasmid DNA)
12. 10 mg/mL tRNA in ddH$_2$O (any tRNA).
13. The DNA fragment to be sequenced, end-labeled at one end only, in ddH$_2$O.
14. 10% piperidine (dilution prepared on the day of the experiment).
15. Loading buffer: 100% formamide, 0.1% (w/v) bromophenol blue, 0.1% (w/v) xylene cyanol FF.

3. Method

1. For each DNA fragment to be sequenced, mark four 1.5-mL Eppendorf "reaction" tubes and add the following solutions:
   "G" tube: 1 µL carrier DNA, 200 µL G Buffer, 5 µL labeled DNA.
   "GA" tube: 1 µL carrier DNA, 10 µL ddH$_2$O, 10 µL labeled DNA.
   "CT" tube: 1 µL carrier DNA, 10 µL ddH$_2$O, 10 µL labeled DNA.
   "C" tube: 1 µL carrier DNA, 15 µL 5M NaCl, 5 µL labeled DNA.
2. Mark up four 1.5-mL Eppendorf "stop" tubes and add the following solutions:
   "G Stop" tube: 2 µL tRNA, 50 µL "G Stop" solution, 1 mL ethanol.
   "AG Stop" tube: 2 µL tRNA, 200 µL "AG Stop" solution, 1 mL ethanol.
   "CT Stop" tube: 2 µL tRNA, 200 µL "CT/C Stop" solution, 1 mL ethanol.
   "C Stop" tube: 2 µL tRNA, 200 µL "CT/C Stop" solution, 1 mL ethanol.
3. To start the reactions (*see* Note 2), add the following:
   "G" tube: 1 µL DMS, mix, and let the reaction proceed for 5 min at RT (*see* Notes 1 and 4).
   "AG" tube: 3 µL 10% formic acid and mix (15 min at 37°C) (*see* Note 4).
   "CT" tube: 30 µL hydrazine and mix (9 min at RT) (*see* Notes 1 and 4).
   "C" tube: 30 µL hydrazine and mix (11 min at RT) (*see* Notes 1 and 4).

4. Stop each reaction by pipetting the contents of the corresponding stop tube into the reaction tube (use the same Pasteur pipet; slight crosscontamination of stop solutions has no effect, but do not touch the contents of the reaction solutions with the pipet). Cap the reaction tubes, shake briefly but vigorously, and place in a dry ice-ethanol bath (–80°C) for 3–10 min (3 min are enough, but the tubes can be left there for up to 10 min if other reactions are not done yet; do not leave for longer than 10 min) (*see* Note 3).

5. Centrifuge at 4°C for 7 min, discard supernatant, aspirate the rest of the liquid with a drawn Pasteur pipet, and then add 1 mL of 100% ethanol to the tube, invert twice, and centrifuge for 2 min at RT. Aspirate as before, and dry in a vacuum for 10 min.

6. To each reaction tube, add 100 µL of the 10% piperidine solution (do not shake the tubes as there is no need to resuspend the DNA) and place the uncapped tubes in a 90°C heat-block. After 15–30 s, cap the tubes and let stand for 30 min.

7. Remove the tubes from the heat-block, let stand at RT for 2–5 min, centrifuge briefly to get the condensation to the bottom. Puncture one hole in the cap with a syringe, then place in dry ice-ethanol bath for 5 min.

8. Place the tubes in a SpeedVac machine and lyophilize for 2 h. Vacuum should be below 100 millitorr.

9. Prior to gel electrophoresis, add 10 µL of loading buffer to each tube, resuspend the sample by shaking and then a brief centrifugation, and place the tubes in the 90°C heat-block for 10 min. Load 1–2 µL per sample (*see* Note 5).

4. Notes

1. Quality of chemicals: In general, the standard laboratory grade chemicals should be used. Some chemicals, however, could be the cause of problems when not sufficiently pure or when too old (presumably degradation products are the culprits). We have only had problems with two chemicals (*see* below): dimethyl sulfate and hydrazine. Note also that these two chemicals, together with piperidine, are hazardous chemicals. In addition to observing the rules pertaining to the handling of radioactive chemicals, all reactions involving these three chemicals should be carried out in the hood.

2. Reaction times: We typically do all chemical reactions together, timing them so that they *end* at the same time. If one is sequencing 5 different fragments, all 20 tubes can be spun together in a single run (we have a microcentrifuge with 20 slots). When stopping all 20 reactions at about the same time, some reactions are invariably going to run longer than the allotted time. This is usually not a problem, because the reaction times indicated above are general, and they can be extended by up to 30% without much noticeable effect.

3. Precipitation: We always use 100% ethanol. There is no need to use any other concentration of ethanol at any step of the process, and 100% ethanol has the advantage because it evaporates fast. The goal is to get the DNA to precipitate with as little salt coprecipitation as possible. This is accomplished by aspirating all the liquid after the ethanol precipitation step, and again after the ethanol wash step. The pellet forms nicely on the side of the tube, and it is easy to put the end of the stretched Pasteur pipet all the way to the bottom of the tube and aspirate all

the liquid. Repeated cycles of resuspension and precipitations are *inadvisable*. Prolonged incubation in the dry ice-ethanol bath is also strongly discouraged. DNA precipitates well at RT, but one gets more salt precipitation at low temperature, thus making things worse, not better. We almost always precipitate DNA at RT; the only reason Step 4 calls for incubation at –80°C is to inhibit further reaction with the reactive reagents, that at this stage have not yet been removed.

4. Troubleshooting:
   - G reaction: This reaction is usually very clean, but it is the reaction most sensitive to prolonged incubation and to the quality of the reactive reagent, DMS. If reaction proceeds longer than the allotted time or if old or bad quality dimethyl sulfate is used, excessive and nonspecific degradation of DNA will occur. Also, when several Gs occur in a row, the 3'-most Gs (lower bands if the 3' end was labeled with the Klenow enzyme) may appear weaker.
   - AG reaction: This is usually a trouble-free reaction.
   - CT and C reactions: The quality of the hydrazine should be good (it does not have to be exceptional), otherwise excessive and nonspecific degradation will occur. Sometimes faint bands will be seen in the C and CT lanes when the base is G (a strong band is then observed in the G lane). The likely explanation is that the pH in the reaction tubes is too low (there is no buffer in the C and CT reaction tubes, but carryover with the DNA sample might cause this to happen). However, these faint bands are not nearly as strong as the signal in the G lane or as bona fide bands of C and T bases. Also, the T bands in the CT lane are often not as strong as the C bands—this is probably caused by inhibition of the reaction by residual salt (in the C reaction, salt is added specifically to obtain complete inhibition).

5. Gel electrophoresis: We use a 60 × 40 cm (0.3-mm thick) gel of 6% acrylamide (20:1 acrylamide:bisacrylamide, 50% urea, 50 mM Tris, 50 mM borate, 1 mM EDTA *[3]*). We run the gel at constant power (65 W), with an aluminum plate to disperse the heat. The samples are loaded twice (a "long run" and a "short run"): The second loading is done when the xylene cyanol dye of the first sample is approx two thirds of the way down the gel, then electrophoresis is halted when the bromophenol blue of the second sample reaches the bottom of the gel. The complete run takes 5–6 h, and it allows us to read, in the "short run," the sequence from about nucleotide 25 to nucleotide 120–150, and in the "long run" the sequence from nucleotide 100 to about 220–250. Additional sequence may be obtained by running longer gels, by loading the sample a third time, or by a variety of other methods if so desired.

References

1. Maxam, A. M. and Gilbert, W. (1980) Sequencing end-labeled DNA with base-specific chemical cleavages. *Methods Enzymol.* **65,** 499–560.
2. Sanger, F., Nicklen, S., and Coulson, A. R. (1977) DNA sequencing with chain-terminating inhibitors. *Proc. Natl. Acad. Sci. USA* **74,** 5463–5467.
3. Maniatis, T., Fritsch, E. F., and Sambrook, J. (1982) *Molecular Cloning, A Laboratory Manual.* Cold Spring Harbor Laboratory, Cold Spring Harbor, NY.

18

Direct PCR Sequencing
with Denaturants (Formamide)

Wei Zhang and Albert B. Deisseroth

1. Introduction

The advance of *Taq*-based polymerase chain reaction (PCR) technology *(1–3)* has had a tremendously positive impact on biomedical research. The combination of PCR and sequencing further revolutionized biological research *(3,4)*. Sequence amplification technology has provided a speedy and effective alternative to work-intensive gene cloning strategies. In addition, small quantities of precious clinical samples that were clearly insufficient for definitive structural analysis by conventional cloning techniques are now sufficient for such analysis using PCR cloning and sequencing methods *(5,6)*. Using these techniques, the genetic changes that exert a dominant effect on the phenotype of neoplastic cells have been brought to light during the past five years *(7–9)*. In addition, many novel genes that govern the proliferation and differentiation of cells have been identified and cloned using redundant primers *(10,11)*. These achievements have greatly advanced our understanding of the regulatory machinery that governs cell growth. The understanding of how these changes in cellular regulation contribute to tumorigenesis has opened the doors to the development of genetic therapy for human diseases.

There are two general approaches for sequencing the DNA products amplified by PCR from cDNA or genomic DNA templates. First, the PCR products can be cloned into a vector, and then the DNA is sequenced by conventional plasmid sequencing. A problem with this approach is that each clone to be sequenced represents only one molecule in a large population of PCR products, thus, the sequence result of any clone can represent a nonrepresentation sampling of a heterogeneous population of cells composed of both normal and tumor cells. In addition, mutations may arise from incorporation errors that

From: *Methods in Molecular Biology, Vol. 65: PCR Sequencing Protocols*
Edited by: R. Rapley Humana Press Inc., Totowa, NJ

occur during the PCR amplification process, although the performance profile of DNA polymerases has been improving as recombinant polymerases with increasing levels of fidelity have been developed each year *(12,13)*. To overcome the problems arising from sequence amplification errors in single clone, multiple clones are normally sequenced *(1–3)*. This approach is therefore tedious, time-consuming, and subjective. Second, the alternative approach is that the PCR product is sequenced directly. The advantage of this latter approach is that the sequencing results represent a population average of the PCR products, and any artifacts derived from misincorporation events generated by the *Taq* DNA polymerase are averaged out. This is owing to the fact that each nucleotide misincorporation event occurs at a single and different position, and therefore represents a low percentage of molecules that contribute to the sequencing ladder.

This approach is straightforward and less time-consuming than the first approach. However, sequencing of a PCR product is quite different technically from sequencing plasmid DNA. PCR sequencing is associated with a set of unique problems, which, if not dealt with decisively, produce tremendous limitations on the results achievable. The first problem is that the PCR fragments are not part of a circular molecule that can be stabilized by plasmid supercoiling. Therefore, the two free complementary strands of PCR fragments have a strong reannealing possibility that the primer-binding efficiency is very low after denaturation. This results in very weak sequencing ladders. The second problem is that PCR fragments with GC-rich regions have an even stronger tendency to reanneal and form secondary structures. This results in high background, weak specific sequencing ladders, ambiguous signals, and premature stops *(14)*. The third problem is that the primers used for PCR appear to have poorer efficiency in binding to the templates under conditions for sequencing. This also results in poor sequencing ladders, as noted above *(15)*.

Many techniques have been developed over the last five years to tackle these problems, and these techniques are described in various chapters in this book. For example, to decrease the reannealing of the complementary PCR fragments and to enhance the binding of primers to templates selectively, "snap-cooling" DNA on ice after boiling was adopted in the presence of an excess amount of primers *(16)*. A rapid decrease in the temperature after denaturation to 4°C inhibits the reannealing of the two complementary PCR strands. The much smaller primer, however, which is in molar excess, has a higher mobility and preferably binds to the complementary template. To overcome the inefficiency of annealing of PCR primers to the templates, "nested" primers are used *(17)*. In this chapter, we will focus on another approach that improves the sequencing of GC-rich PCR fragments and PCR sequencing using PCR primers *(14,15)*.

Our analyses have led us to conclude that one way to solve the problems associated with the reannealing of PCR complementary strands is to decrease the probability of formation of hydrogen bonds between the two strands of PCR fragments, especially the GC-rich fragments. It has been known for a long time that DNA denaturants can accomplish this, and there are two denaturants commonly used for this purpose, formamide *(18,19)* and DMSO *(20,21)*. Our studies showed that formamide could markedly improve the quality or the sequencing results either with PCR templates, which have GC-rich regions, or when the same PCR primers are used for sequencing *(14,15)*. The use of DMSO also improves the quality of PCR sequencing results in a similar manner *(22)*. The protocol presented here is an easy and effective way to obtain such satisfactory sequencing results. It does not require expensive or hard-to-get chemical reagents, and it only requires small amounts of PCR fragments (50–200 ng). Finally, it works well with PCR fragments in the size range of 110 bp to 1.2 kb *(see* Note 1).

2. Materials and Equipment

The reagents used should be molecular-biology-grade, and double-distilled water should be used to make all buffers used in the experiments. The reagents for sequencing are available in the Sequenase kit from US Biochemicals (Cleveland, OH), Amersham (Arlington Heights, IL).

2.1. Purification of PCR Products from Polyacrylamide Gel

1. Polyacrylamide and bis-acrylamide: These reagents are neurotoxic. Masks are recommended when preparing the solution from powder.
2. Tris-boric acid-EDTA (TBE) buffer *(see* Note 2).
3. A vertical gel apparatus (model V16, BRL, Gaithersburg, MD): 1.2-mm thick spacer.
4. 6X DNA gel-loading dye: 0.25% bromophenol blue, 0.25 xylene cyanol FF, 30% glycerol in water.
5. Spectrum dialysis tubing *(see* Note 3) and clamps.
6. A horizontal gel apparatus.
7. A power supply.
8. $3M$ sodium acetate, pH 4.6–6.0.
9. 100% Ethanol.

2.2. Sequencing of the PCR Products

1. 5X Sequenase buffer: 200 mM Tris-HCl, pH 7.5, 100 mM MgCl$_2$, 250 mM NaCl.
2. Formamide: Store at 4°C.
3. 0.1M DTT: Store at −20°C.
4. Labeling mix: A 7.5-μM solution of each deoxynucleotide triphosphate (dNTP). Store at −20°C. Dilute eightfold in water for use in sequencing.
5. ^{35}S-dATP (sequencing-grade, Du Pont): This is a low-energy β-emitter; no screen is needed. However, all waste products should be stored in designated containers and disposed of following institutional policy. The half-life of this isotope is about 3 mo.
6. Sequenase dilution buffer: 10 mM Tris-HCl, pH 7.5, 5 mM DTT, 0.5 mg/mL BSA.

7. Sequenase (Version 2.0, US Biochemicals): Should be stored at –20°C at all times. Keep on ice and dilute eightfold with Sequenase dilution buffer shortly before use.
8. Termination mixes: mixture of 80 μM dNTP plus 8 μM of ddATP for A termination mix, 8 μM ddTTP for T termination mix, 8 μM ddCTP for C termination mix, and 8 μM ddGTP for G termination mix, in 50 mM NaCl.
9. Stop solution: 95% formamide, 20 mM EDTA, 0.05% bromophenol blue, 0.05% xylene cyanol.
10. Sequencing gel: A 0.4-nm thick, 6% denaturing polyacrylamide gel (*see* Note 4).
11. Autoradiographic film: Kodak XAR5 film.
12. Cassette: No intensifying screens are needed.

3. Methods

3.1. Purification of PCR Fragments from Acrylamide Gel by Electroelution

We routinely purify the PCR product from a 5% nondenaturing acrylamide gel. PCR primers, free nucleotides, and other contaminants can be eliminated during this procedure, and the resulting DNA is very clean for satisfactory sequencing. PCR products can also be purified by commercially available spin columns, especially if the PCR reaction produces a very specified product (one clear band on the gel). However, it has been noticed that some spin columns may have inhibitors to Sequenase; therefore, tests are needed to determine this. PCR products purified from acrylamide gels give consistent results.

1. Prepare a 5% nondenaturing polyacrylamide gel (0.8-mm thick, TBE buffer) in a vertical gel apparatus (same apparatus used for protein gel); use a comb with wide teeth.
2. Pool the DNA mixtures from several PCR reactions (depending on the yield of PCR reaction). Add 1/5 vol of the 6X DNA gel-loading buffer. Electrophorese at 200 V for several hours, depending on the size of the PCR products.
3. Following conclusion of the electrophoresis, stain the gel with ethidium bromide briefly. Locate the band with long-wave, hand-held UV light box. Cut out the band with a razor blade.
4. Fill a precleaned and soaked dialysis tubing with water, and slip the DNA band into the tubing. Remove most of the water, leaving about 500 μL of water in the tube. Squeeze out the air bubbles, and clamp the tube. Move the gel slice to one side of the tube by hand (*see* Fig. 1).
5. Place the tubing on a horizontal electrophoresis box (for routine DNA agarose gel electrophoresis) filled with 0.1X TBE. The gel side of the tubing is toward the negative electrode. Electrophorese at 200 V for 30 min. The DNA should be eluted into the buffer in the tubing. The completion of electrophoresis can be easily checked by the hand-held UV light box.
6. Remove the DNA in water by pipet, and transfer into a 1.5-mL Eppendorf tube. Centrifuge at full speed (14,000*g*) for 10 min to remove gel debris.

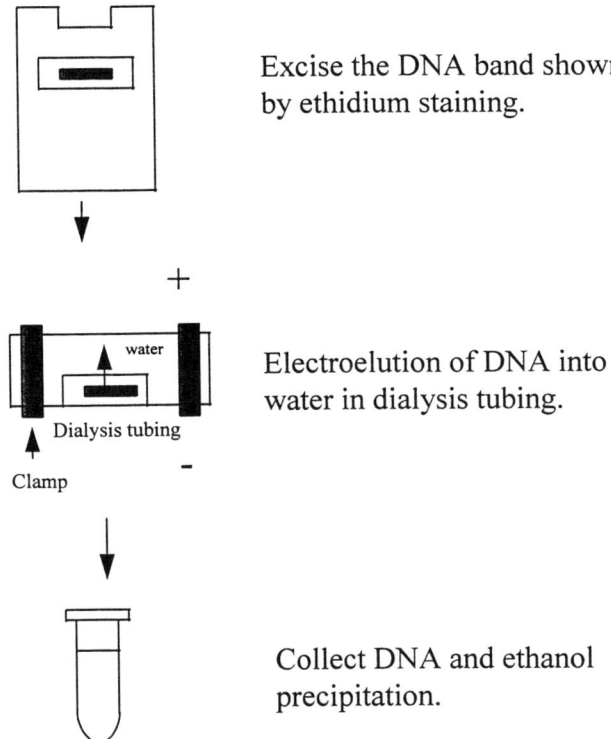

Excise the DNA band shown by ethidium staining.

+

Electroelution of DNA into water in dialysis tubing.

Collect DNA and ethanol precipitation.

Fig. 1. Purification of PCR products from polyacrylamide gel by electroelution.

7. Transfer the supernatant into a new Eppendorf tube, add 1/10 vol of 3*M* NaOAc, and 2 vol of cold ethanol. Mix well and precipitate the DNA for 15 min at −70°C. Spin down the DNA into a pellet, and wash the pellet twice with 70% ethanol. Dry the DNA in a SpeedVac, and resuspend the DNA into appropriate volume.
8. Estimate the concentration of DNA by spectrophotometry or by comparative staining with ethidium bromide.

3.2. Sequencing of PCR Products with the Aid of Formamide

We use the following procedure to sequence PCR fragments having GC-rich regions, or we use the PCR primers for direct sequencing. The majority of reagents used in this procedure are from the US Biochemicals Sequenase kit:

1. Mix 50–200 ng of PCR fragments and a fivefold molar excess of primers in a 10 μL 1X Sequenase buffer containing 10% formamide (*see* Notes 5–8). Then boil the mixture for 8 min to denature the PCR fragments.
2. Quickly put the Eppendorf tube on ice or on a dry ice–ethanol bath to fast-cool the reaction (*see* Note 9). Leave the reactions on ice or on a dry ice–ethanol bath for 2 min.

3. Spin the tubes briefly at 4°C to collect all the liquid on the bottom of the tubes (*see* Note 10). Transfer the tubes onto ice.
4. Add 1 µL of 0.1*M* DTT, 2 µL of diluted labeling mix, 0.5 µL of ^{35}S-dATP, and 2 µL of freshly diluted Sequenase. Incubate on ice for 10 min (*see* Note 11).
5. A set of four tubes should be prepared marked "G," "A," "T," and "C." 2.5 µL of ddGTP, ddATP, ddTTP, and ddCTP are added to one of the four tubes, respectively. Prewarm the tubes to 42–45°C for 1 min before the finish of reaction at step 4 (*see* Note 12).
6. Transfer 3.5 µL of reaction mixture to each of the prewarmed tubes, and continue the incubation for 5 min.
7. Add 1 µL of diluted labeling mix to each of the tubes. Incubate for an additional 5 min (*see* Note 13).
8. Add 4 µL of stop solution to the reaction. Mix well and leave on ice.
9. Prerun the sequencing gel (BRL model S2) for 30 min with constant power of 55 W (*see* Notes 14 and 15).
10. Heat the samples at 95°C for 3 min. Meanwhile, stop the prerun and wash the wells four to five times using a syringe (*see* Note 16).
11. Load 2.5–3 µL of reaction mixture (*see* Note 17). The commonly used loading patterns are CATG and GATC. Continue to run the gel at 55 W.
12. After the gel run is complete, transfer the gel onto a piece of Whatman 3MM paper. Cover the gel with a film wrap, and dry the gel at 80°C under vacuum (*see* Chapter 1).
13. Place a sheet of X-AR film on the top of the gel in a cassette (*see* Note 18). Autoradiograph at –70°C overnight. Allow the cassette to warm up and dry the following day. Develop the film in an automatic film developer. Examples of sequencing results are shown in Figs. 2 and 3.

4. Notes

1. Use 200 ng for a 1.2-kb DNA fragment; smaller amounts of DNA can be used for smaller-size DNA fragments.
2. Precipitates form from TBE buffer with time; therefore, make new buffer when this happens. Use a new clear glass bottle each time.
3. The molecular cutoff sizes do not appear to matter, because DNA is linear and not in a globular form. It is, therefore, too big to pass through the pores of the dialysis tubing.
4. We found that the gel made fresh from powder gives the best results for the sequencing ladder. Urea, acrylamide, and bis-acrylamide should be dissolved into 0.5–1X TBE at 37°C in a water bath. Ammonium persulfate (APS) should then be added and dissolved. This solution should then be vacuum filtered for 5 min to remove undissolved particles and air bubbles. TEMED is then added and mixed gently by pipeting; however, bubbles should be avoided. After the solution is poured, the gel needs to be polymerized for 2 h before use. The recipe for the 6% gel (0.5X TBE, 75 mL) is the following: 3.7 mL 10X TBE, 37.5 g urea, 4.5 g acrylamide, 0.23 g bis-acrylamide, 0.06 g APS, 33 mL water, and 20 µL TEMED.

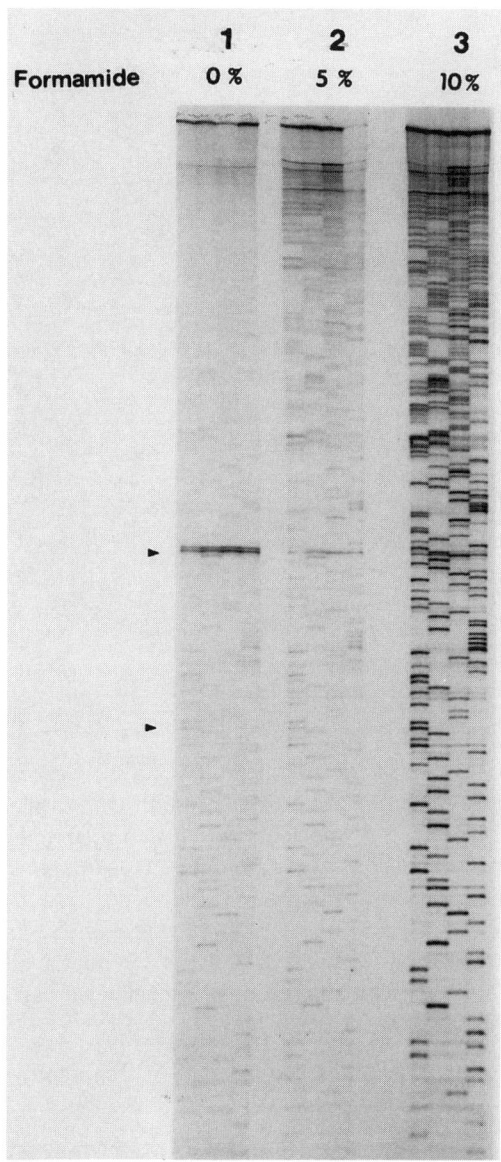

Fig. 2. Two hundred-nanogram PCR products of *p53* gene (1.3 kb) were mixed with a fivefold molar excess of sequencing primer in Sequenase buffer with the increasing percentages of formamide indicated in each lane. The annealing mixture was boiled for 10 min and quickly cooled in a dry ice–ethanol bath. Sequencing reactions were carried out on ice for 10 min, then aliquoted to prewarmed ddNTP, and reacted for 10 min at 42°C. The reaction was stopped by adding stop solution. The sequencing reactions were heated to 95°C for 3 min before loading. The loading pattern was GATC. The region being sequenced contains 62% GC.

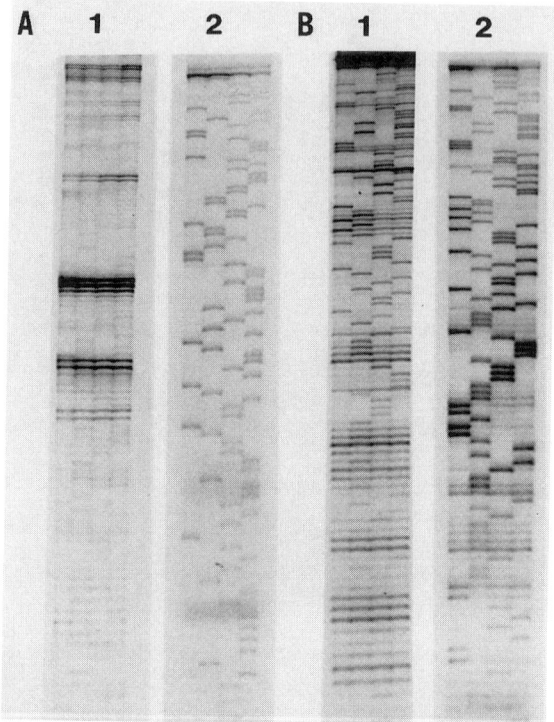

Fig. 3. Fifty nanograms of PCR fragments (≈120 bp) were mixed with a fivefold molar excess of primers in Sequenase buffer containing no formamide (A1 and B1) or 10% formamide (A2 and B2). A and B represent two different PCR fragments with 50% GC content, sequenced using two different PCR primers.

5. We routinely use 200 ng for DNA fragments of about 1 kb and 50 ng for fragments of about 200 bp. The amount of primer needed can be calculated by the following formula:

$$\mu g \text{ of primer} = 5 \times \mu g \text{ of PCR fragment} \times \text{size of primer} \div \text{size of PCR DNA} \quad (1)$$

A tenfold molar excess of primer can also be used in cases in which insufficient annealing is observed.

6. The role of formamide is to weaken the hydrogen bonds between the two strands of the PCR fragment, thus reducing the secondary structure that causes the ambiguity in banding pattern. The percentage of formamide required to generate the improvement of sequencing depends on the percentage of GC nucleotides in the target sequence. It may therefore be necessary to determine empirically the amount of formamide needed to resolve secondary structures completely. The following formula can be used to calculate the percentage of formamide needed to

reduce the melting temperature of the DNA duplex to that of a fragment with a 50% GC nucleotide:

$$\text{Formamide \%} = 0.7\,(\text{GC\%} - 50) \tag{2}$$

An example of the effect of formamide is shown in Fig. 2. It should be pointed out that some nonspecific bands of secondary structure cannot be totally abolished, even with 30% formamide because formamide reduces the efficiency of the findings of the template to primer binding, especially when the GC percentage of the primer is much lower than the GC percentage of the local structure.

7. Formamide (10%) is routinely used when the PCR primer is used to sequence the PCR product. The result is much improved even for non-GC-rich fragments (*see* Fig. 3).

8. DMSO can be used in place of formamide. The percentage of DMSO needs to be determined empirically.

9. When the dry ice–ethanol bath is used, do not have flowing ethanol, because the ethanol may get inside the tube.

10. This is needed because the solution evaporates during boiling.

11. The reaction is carried out on ice to limit the reannealing of two complementary strands from PCR products.

12. The temperature of 42–45°C is used instead of 37°C to help overcome the secondary structure.

13. This step helps force the unfinished reaction to completion, thereby reducing premature stops that may cause bands across all four lanes.

14. The TBE buffer used for the running gels should be consistent with the buffer used in the gel. For example, if 0.5X TBE was used to make the gel, 0.5X TBE should be used to run the gel. For an extended long run, 1X TBE is preferred.

15. The gel should be warm (about 50°C) after the prerun. If the plate is not warm, the wrong buffer may have been used.

16. Be careful not to shift the comb by the needle. Also, be careful not to wash the wells too forcefully.

17. If a small leak of sample occurs during loading, start the running of the electrophoresis after loading one sample set. Let the DNA run into the gel, and then load the next set or sets.

18. On many occasions, gels do not completely dry. If this happens, do not peel off the film wrap used on top of the gel so that the X-AR film will not get stuck to the gel. However, if the gel is completely dry, the film wrap can be peeled off so that stronger signals can be obtained.

Acknowledgment

We thank Leslie Wildrick for her editorial assistance.

References

1. Saiki, R. K., Scharf, S., Faloona, F., Mullis, K. B., Horn, G. T., Erlich, H. A., and Arnheim, N. (1985) Enzymatic amplification of β-globin genomic sequences and restriction site analysis for diagnosis of sickle cell anemia. *Science* **230**, 1350–1354.

2. Mullis, K. B. and Faloona, F. A. (1987) Specific synthesis of DNA in vitro via a polymerase-catalyzed chain reaction. *Methods Enzymol.* **155,** 335–350.
3. Scharf, S. J., Horn, G. T., and Ehrlich, H. A. (1986) Direct cloning and sequence analysis of enzymatically amplified genomic sequences. *Science* **233,** 1076–1078.
4. Innis, M. A., Myambo, K. B., Gelfand, D. H., and Brow, M. A. D. (1988) DNA sequencing with *Thermus aquaticus* polymerase and direct sequencing of polymerase chain reaction-amplified DNA. *Proc. Natl. Acad. Sci. USA* **85,** 9436–9440.
5. Sarkar, F. H., Li, Y. W., and Crissman, J. D. (1993) A method for PCR sequencing of the pT53 gene from a single 10-micron frozen or paraffin embedded tissue section. *BioTechniques* **15,** 36–38.
6. Trumper, L. H., Brady, G., Bagg, A., Gray, D., Loke, S. L., Griesser, H., Wagman, R., Gascoyne, R. D., and Vicini, S. (1993) Single-cell analysis of Hodgkin and Reed-Sternberg cells: molecular heterogeneity of gene expression and p53 mutations. *Blood* **81,** 3097–3115.
7. Wong, C., Dowling, C. E., Saiki, R. K., Higuchi, R. G., Ehrlich, H. A., and Kazazian, H. H. (1987) Characterization of β-thalassemia mutations using direct genomic sequencing of amplified genomic DNA. *Nature* **330,** 384.
8. Baker, S. J., Fearon, E. R., Nigro, J. M., Hamilton, S. R., Preising, A. C., Jessup, J. M., van Tuinen, P., Ledbetter, D. H., Barker, D. F., Nakamura, Y., White, R., and Vogelstein, B. (1989) Chromosome 17 deletions and p53 gene mutations in colorectal carcinomas. *Science* **244,** 217–221.
9. Hu, G., Zhang, W., and Deisseroth, A. B. (1992) p53 gene mutations in acute myeloid leukemia. *Br. J. Haematol.* **81,** 489–494.
10. Xiong, Y., Hannon, G., Zhang, H., Casso, D., Kobayashi, R., and Beach, D. (1993) p21 is a universal inhibitor of cyclin kinases. *Nature* **366,** 701–704.
11. Scarpetta, M. A. and Uhler, M. D. (1993) Evidence for two isoforms of the endogenous protein kinase inhibitor of cAMP-dependent protein kinase in mouse. *J. Biol. Chem.* **268,** 10,927–10,931.
12. Barnes, W. M. (1992) The fidelity of Taq polymerase catalyzing PCR is improved by an N-terminal deletion. *Gene* **112,** 29–35.
13. Cheng, S., Fockler, C., Barnes, W. M., and Higuchi, R. (1994) Effective amplification of long targets from cloned inserts and human genomic DNA. *Proc. Natl. Acad. Sci. USA* **91,** 5695–5699.
14. Zhang, W., Hu, G., and Deisseroth, A. B. (1991) Improvement of PCR sequencing by formamide. *Nucleic Acids Res.* **19,** 6649.
15. Zhang, W., Reading, C., and Deisseroth, A. B. (1992) Improved PCR sequencing with formamide. *Trends Genet.* **8,** 332.
16. Kusukawa, N., Uemori, T., Asada, K., and Kato, I. (1990) Rapid and reliable protocol for direct sequencing of material amplified by the polymerase chain reaction. *BioTechniques* **9,** 66–72.
17. Phear, G. A. and Harwood, J. (1994) Direct sequencing of PCR products, in *Methods in Molecular Biology* (Harwood, A. J., ed.), Humana, Totowa, NJ, pp. 247–256.

18. Casey, J. and Davidson, N. (1977) Rates of formation and thermal stabilities of RNA:DNA and DNA:DNA duplexes at high concentrations of formamide. *Nucleic Acids Res.* **4,** 1539–1552.
19. McConaughy, B. L., Laird, C. D., and McCarthy, B. J. (1969) Nucleic acid reassociation in formamide. *Biochemistry* **8,** 3289–3295.
20. Chester, N. and Marshak, D. R. (1993) Dimethyl sulfoxide-mediated primer T_m reduction: a method for analyzing the role of renaturation temperature in the polymerase chain reaction. *Anal. Biochem.* **209,** 284-290.
21. Masoud, S. A., Johnson, L. B., and White, F. F. (1992) The sequence within two primers influences the optimum concentration of dimethyl sulfoxide in the PCR. *PCR Methods Appl.* **2,** 89–90.
22. Winship, P. R. (1989) An improved method for directly sequencing PCR amplified material using dimethyl sulphoxide. *Nucleic Acids Res.* **17,** 1266.

19

Efficient PCR Production
of Single-Stranded DNA Sequencing Templates

Bernhard Kaltenboeck and Konstantin G. Kousoulas

1. Introduction

The acceptance of polymerase chain reaction (PCR) as a routine method in molecular biology has created a need for simple and robust approaches to DNA sequencing of the amplification products. In addition to its ease, direct sequencing of PCR products has the advantage that nucleotide misincorporation during amplification does not pose a problem, because only a small percentage of the DNA molecules may contain such random mutations. However, a large body of published protocols attests to the difficulties encountered in this endeavor. Double-stranded sequencing of the linear PCR products is notoriously unreliable because of rapid reannealing of the denatured DNA.

This chapter describes a simple and effective PCR method to produce single-stranded (ss) DNA for high-quality sequencing directly from double-stranded (ds) amplification products. Gyllensten and Erlich *(1)* first conceived a modified PCR, termed asymmetric PCR, to generate ssDNA for sequencing. Using skewed primer concentrations, this PCR initially amplifies dsDNA, which then serves as template for linear, run-off production of ssDNA during the last thermal cycles, when the limiting primer has been exhausted. Inconsistent results prompted several groups to separate the asymmetric amplification into ds and ssPCRs: dsDNA amplified in the first reaction was the starting material for production of ssDNA in a second PCR with one primer in excess *(2–8)*.

We analyzed the principles underlying the successful production of high-quality ssDNA in this dual PCR approach, and found that the factors of overriding importance were primer concentrations in ds and ssPCR, dsPCR input into the ssPCR, and the number of cycles of the ssPCR *(5)*. If the dsPCR uses primers at 0.2 μM each, then the optimum concentration of the single primer in

From: *Methods in Molecular Biology, Vol. 65: PCR Sequencing Protocols*
Edited by: R. Rapley Humana Press Inc., Totowa, NJ

a 100-μL ssPCR is 0.4 μ*M* with a direct input of 10 μL of the dsPCR. The optimum number of ssPCR cycles is generally between 15 and 20, and needs to be determined for each primer.

The 10-μL dsPCR input typically contains 50–100 ng of the dsDNA template and ~2 pmol of each primer. The large amount of dsDNA allows for effective run-off production of ssDNA in few thermal cycles. In addition, the 2 pmol of limiting primer carried over from the dsPCR mediate a "fill-up" dsDNA amplification, resulting in a self-adjustment of the dsDNA template concentration.

The low number of ssPCR cycles is critical because it prevents the exponential buildup of aberrant DNA fragments primed by the single primer. Protocols calling for more thermal cycles in the ssPCR are notoriously unreliable, because the reactions become transformed from linear primer extensions into exponential amplifications of these aberrant products.

We have successfully applied these principles to numerous PCRs employing different primers and reaction parameters, and have obtained unequivocal sequencing ladders from these ssDNA products (5). We believe virtually any DNA fragment that can be amplified by PCR would be amenable to this type of ssDNA production and sequencing using any of the established techniques for ssDNA sequencing.

2. Materials

1. 10X PCR buffer: 200 m*M* Tris-HCl, pH 8.3, 250 m*M* KCl, 20 m*M* MgCl$_2$, 0.5% Tween-20, 0.5% Nonidet P-40; autoclave and store in small quantities at −20°C.
2. 10 mg/mL Bovine serum albumin (BSA): molecular-biology-grade.
3. Deoxyribonucleotide triphosphate mixture (dNTPs; mixture of 2 m*M* each of dATP, dCTP, dGTP, and dTTP).
4. Upstream and downstream oligonucleotide primers, all 2 μ*M*.
5. 5 U/μL *Taq* DNA polymerase.
6. Mineral oil.
7. Centricon-100® centrifugal ultrafiltration devices (Amicon, Beverly, MA).

3. Methods

3.1. dsPCR (see *Note 1*)

1. Assemble a 100-μL dsPCR at room temperature in a thermocycling tube:
   a. 48.5 μL of H$_2$O (deionized, autoclaved).
   b. 10 μL of 10X PCR buffer (maintain at room temperature, not on ice).
   c. 1 μL of BSA.
   d. 10 μL of dNTPs.
   e. 10 μL of Upstream primer.
   f. 10 μL of Downstream primer.
   g. 0.5 μL of *Taq* DNA polymerase.
   h. 10 μL of Target DNA solution.

2. Mix by tapping with finger, layer two drops of mineral oil on top of reaction, briefly spin down, and place into thermocycler.
3. Thermocycle the PCR mix using a protocol appropriate to yield a single, amplified target DNA fragment.
4. Remove 10 μL of sample from under the oil and examine on a 2% agarose gel.
5. Proceed to ssPCR, if amplification of one discrete dsDNA fragment has been accomplished. If not, optimize the PCR before producing ssDNA for sequencing.

3.2. Optimization of ssPCR (see Notes 2 and 3)

1. To determine for each primer the optimal number of ssPCR cycles, assemble a 100-μL ssPCR at room temperature:
   a. 48.5 μL of H$_2$O (deionized, autoclaved).
   b. 10 μL of 10X PCR buffer (maintain at room temperature, not on ice).
   c. 1 μL of BSA.
   d. 10 μL of dNTPs.
   e. 20 μL of either upstream (for sense ssDNA) or downstream (for antisense ssDNA) primer.
   f. 2.5 U/0.5 μL of *Taq* DNA polymerase.
   g. 10 μL of dsPCR sample.
2. Mix by tapping with finger, layer two drops of mineral oil on top of reaction, briefly spin down, and incubate on the thermocycler.
3. Perform a 30-cycle reaction using a modified thermocycling protocol established for dsPCR: the primer extension segment at 72°C should be at least 2 min.
4. Remove 10-μL samples each of the ssPCRs after 10, 15, 20, 25, and 30 thermal cycles from under the oil. Analyze these samples on a 2% agarose gel. The ssDNA fragment stains less intensively with ethidium bromide and generally migrates faster than the dsDNA (except in TBE buffer, where some ssDNA fragments migrate more slowly).
5. The optimum number of ssPCR cycles is, in most cases, between 15 and 20 and should be identified by the following criteria:
   Reactions containing little or no visible DNA fragments except for the ds and ss target DNA are preferable to those with aberrant products even if these contain more of the ssDNA.
   Best sequencing results are obtained with ssPCRs containing the least amount of dsDNA relative to ssDNA.

3.3. Production of ssDNA in ssPCR (see Note 4)

1. Set up a 100-μL ssPCR as described above.
2. Thermocycle for the optimum number of cycles with at least 2 min of primer extension at 72°C.
3. Remove the complete reaction mix from under the oil. Examine a 10-μL portion on a 2% agarose gel.
4. Dilute the remaining 90 μL of the ssPCR in a Centricon-100 device to 1.8 mL with dH$_2$O. Spin according to manufacturer's instructions to remove buffer, primers, and dNTPs.

5. Fill sample chamber of the Centricon-100 device with 2 mL dH$_2$O and spin again. Repeat this step once.
6. Recover ultrafiltrated retentate (~45 µL) in retentate cup by centrifuging the inverted Centricon-100 device according to manufacturer's instructions.
7. The ultrafiltrate is ready for use in ssDNA sequencing.

3.4. Direct PCR Sequencing

Perform direct sequencing on ssPCR products as detailed in Chapter 5, Section 3.3. Analyze sequencing products as indicated in Chapter 1.

4. Notes

1. It is of great importance that the dsPCR be optimized to yield a single, discrete target DNA fragment. Suitable PCR parameters and steps to optimize them are reviewed by Bangham *(9)* and Innis and Gelfand *(10)*.

   We routinely use the dsPCR mix as described and rarely find it necessary to optimize new PCRs, except for the precise determination of the annealing temperature. We prefer primers of ≥20 nucleotides, because the resulting high annealing temperatures allay problems of artifacts caused by nonspecific priming and secondary structure of the primers. The primer concentration of 0.2 µ*M* each is generally applicable. This concentration does not negatively affect yields of the amplification product in PCRs with low DNA background, yet at high DNA background (~1 µg/100 µL PCR), the sensitivity of the PCR is much better. We find the 2 m*M* magnesium ion concentration virtually always ideal, if the dNTP concentration is 0.2 m*M* each. Up to 20-fold reduction of the dNTP concentration sometimes alleviates the problem of nonspecific amplification products, but it also reduces the yield of the PCR. In our hands, some reaction components are critical for high yields in PCR: 0.1% of nonionic detergents like Tween-20 or Nonidet P-40, and 0.01% of BSA instead of gelatin.

2. Primer concentrations other than 0.2 µ*M* in the dsPCR change the optimum dsPCR input into the ssPCR: if, for example, the concentration was 0.5 µ*M*, then only about 4 µL input would be feasible, resulting in less dsDNA template available for ssDNA production. More dsPCR input would bias the ssPCR into dsDNA production owing to a carryover of too much limiting primer *(5)*.

   Conversely, low primer concentrations like 0.1 µ*M* in the dsPCR typically reduce the yield of amplified dsDNA. In many cases, a 20-µL compensatory dsPCR input into the ssPCR cannot offset this disadvantage, if the single primer concentration is not optimized for the ssPCR.

3. The ssPCR extension thermal segment should be long enough (≥2 min/1000 bp of dsDNA fragment) to allow for production of full-length ssDNA. Although numerous parameters influence the production of ssDNA (e.g., single primer and *Taq* DNA polymerase concentrations, annealing temperature), it is not necessary to optimize them. Instead, these minor effects on the ssDNA yield can be compensated for by adjusting the number of ssPCR cycles.

4. For use in Sequenase® (United States Biochemicals, Cleveland, OH) ssDNA sequencing, 7-μL portions of the ultrafiltrate generally are sufficient, if the ssDNA is shorter than 1000 nucleotides (~12% of the ssDNA of a 100 μL ssPCR, equivalent to ~0.3 pmol of the ssDNA). The ultrafiltrate should be concentrated twofold for longer templates to compensate for their lower molarity, and for improved reading of sequences close to the primer in ^{35}S sequencing. Like Gibbs et al. *(3)*, we have obtained better results with performing Sequenase sequencing reactions at 50°C than with reactions incubated at 37°C.

References

1. Gyllensten, U. B. and Erlich, H. A. (1988) Generation of single-stranded DNA by the polymerase chain reaction and its application to direct sequencing of the HLA-DQA locus. *Proc. Natl. Acad. Sci. USA* **85,** 7652–7656.
2. Allard, M. W., Ellsworth, D. L., and Honeycutt, R. L. (1991) The production of single-stranded DNA suitable for sequencing using the polymerase chain reaction. *BioTechniques* **10,** 24–26.
3. Gibbs, R. A., Nguyen, P.-A., and Caskey, C. T. (1991) Direct DNA sequencing of complementary DNA amplified by the polymerase chain reaction, in *Methods in Molecular Biology*, vol. 9, *Protocols in Human Molecular Genetics* (Mathew, C., ed.), Humana, Totowa, NJ, pp. 9–20.
4. Kadowaki, T., Kadowaki, H., and Taylor, S. I. (1990) A nonsense mutation causing decreased levels of insulin receptor mRNA: detection by a simplified technique for direct sequencing of genomic DNA amplified by polymerase chain reaction. *Proc. Natl. Acad. Sci. USA* **87,** 658–662.
5. Kaltenboeck, B., Spatafora, J. W., Zhang, X., Kousoulas, K. G., Blackwell, M., and Storz, J. (1992) Efficient production of single-stranded DNA as long as 2 kb for sequencing of PCR-amplified DNA. *BioTechniques* **12,** 164–171.
6. Landweber, L. F. and Kreitman, M. (1993) Producing single-stranded DNA in polymerase chain reaction for direct genomic sequencing. *Methods Enzymol.* **218,** 17–26.
7. Medori, R., Tritschler, H.-J., and Gambetti, P. (1992) Production of single-stranded DNA for sequencing: an alternative approach. *BioTechniques* **12,** 347–349.
8. Nichols, R. C. and Raben, N. (1994) Hints for direct sequencing of PCR-generated single-stranded DNA. *BioTechniques* **17,** 412–414.
9. Bangham, C. R. M. (1991) The polymerase chain reaction. Getting started, in *Methods in Molecular Biology*, vol. 9, *Protocols in Human Molecular Genetics* (Mathew, C., ed.), Humana, Totowa, NJ, pp. 1–9.
10. Innis, M. A. and Gelfand, D. H. (1990) Optimization of PCRs, in *PCR Protocols. A Guide to Methods and Applications* (Innis, M. A., Gelfand, D. H. Sninsky, J. J., and White, T. J., eds.), Academic, New York, pp. 3–12.

20

Preparation and Direct Automated Cycle Sequencing of PCR Products

Susan E. Daniels

1. Introduction

The polymerase chain reaction (PCR) is well known for being a rapid and versatile method for the amplification of defined target DNA sequences. This technique can be applied to a variety of research areas, such as the identification and typing of single nucleotide substitutions of DNA sequence polymorphisms, and genetic mapping *(1–4)*.

Since the introduction of PCR *(5)*, a variety of methods for sequencing PCR-generated fragments have been described. These are usually based on the Sanger chain-terminating dideoxynucleotide sequencing *(6)*, rather than the Maxam and Gilbert chemical cleavage method *(7)*. Manual dideoxy sequencing methods are labor intensive, time-consuming, involve radioisotopes, and have limitations in sequence ordering. However, a technique combining the PCR and dideoxy terminator chemistry simplifies the process of sequencing and is known as cycle sequencing *(8)*. Automated or fluorescent DNA sequencing is a variation of the traditional Sanger sequencing utilizing the cycle sequencing methodology, where fluorescent labels are covalently attached to the reaction products, and data are collected during the polyacrylamide gel electrophoresis.

The introduction of fluorescently labeled dideoxynucleotides as chain terminators presented the opportunity for the development of reliable cycle sequencing for PCR products. The sequencing reaction with the dye terminators is carried out in a thermal cycler, and each of the four dideoxynucleotide triphosphates (ddNTPs) is labeled with a different fluorescent dye. This allows the four chain extension reactions to be carried out within a single tube, sparing considerable labor *(9,10)*. The use of labeled chain terminators allows flexibility of sequencing strategy, since the same primers can be used in the sequencing reaction.

From: *Methods in Molecular Biology, Vol. 65: PCR Sequencing Protocols*
Edited by: R. Rapley Humana Press Inc., Totowa, NJ

This eliminates the time and expense associated with a separate set of modified DNA sequencing primers and is well suited to high throughput sequencing.

Using this method, it is possible to amplify a target DNA sequence, purify the resulting fragment, and obtain sequencing data within 24 h. Also, it has been used to sequence a 500-bp PCR fragment on an automated DNA sequencer with 99.3% accuracy *(10–12)*.

This chapter will concentrate on the techniques involved with the direct sequencing reactions rather than on the use of the machine, since each automated DNA sequencer will be provided with an extensive manual for its operation.

2. Materials

All solutions should be made to the standard required for molecular biology. Use molecular-biology-grade reagents and sterile distilled water. The reagents for the cycle sequencing are available commercially.

2.1. Purification of PCR Products Before Cycle Sequencing

1. $4M$ Ammonium acetate.
2. Isopropanol.
3. 70% (v/v) Ethanol.
4. 10 mM Tris-HCl, pH 7.5, 1 mM EDTA.

2.2. Cycle Sequencing

Prism™ ready reaction DyeDeoxy terminator premix (1000 µL), Applied Biosystems (ABI) consists of 1.58 mM A-dyedeoxy, 94.74 µM T-dyedeoxy, 0.42 µM G-dyedeoxy, 47.37 µM C-dyedeoxy, 78.95 µM dITP, 15.79 µM dATP, 15.79 µM dCTP, 15.79 µM dTTP, 168.42 mM Tris-HCl (pH 9.0), 4.21 mM $(NH_4)_2SO_4$, 42.1 mM $MgCl_2$, 0.42 U/µL AmpliTaq DNA polymerase.

2.3. Purification of PCR Products After Cycle Sequencing

1. Chloroform.
2. Phenol:H_2O:chloroform (16:18:14) at room temperature.
3. $2M$ Sodium acetate, pH 4.5.
4. 100 and 70% (v/v) ethanol at room temperature.

2.4. 6% Polyacrylamide Sequencing Gels

1. 10X TBE: 890 mM Tris-borate, 890 mM boric acid, 20 mM EDTA, pH 8.3.
2. 40 g Urea.
3. 12 mL 40% (w/v) acrylamide stock solution (19:1 acrylamide/bis-acrylamide).
4. 20 mL dH_2O.
5. 1 g mixed-bed ion-exchange resin.
6. TEMED.
7. 10% (w/v) Ammonium persulfate, freshly made.

2.5. Loading Buffer

1. 50 m*M* EDTA, pH 8.0.
2. Deionized formamide.

3. Methods

3.1. Isopropanol Purification of PCR Products

It is essential to remove excess PCR primers before using DyeDeoxy terminators for cycle sequencing (*see* Note 1).

1. Aliquot an appropriate amount of the PCR reaction into a 0.6-mL microfuge tube, and dilute to a total of 20 μL with distilled water.
2. Add 20 μL 4*M* ammonium acetate into the microfuge tube, and mix well.
3. Add 40 μL of isopropanol into the tube, mix well, leave at room temperature for 10 min, and centrifuge the microfuge tube for 10 min at 12,000*g*.
4. Carefully remove the supernatant, and wash the pellet with 70% (v/v) ethanol. Then briefly dry the pellet under vacuum.
5. Resuspend the pellet in 20 μL of TE buffer.

3.2. Cycle Sequencing of PCR Products

The amount of PCR product should be estimated on an agarose gel before sequencing. Approximately 1 μg of double-stranded DNA template or 0.5 μg of single-stranded DNA template is required for each sequencing reaction.

1. Mix the following reagents in a 0.6-mL microfuge tube, 5-mL DNA template, 1-μL primer (from a 3.2-pmol stock solution), and 4.5 μL sterile dH₂O (*see* Notes 4 and 5).
2. Add 9.5 μL ABI Prism ready reaction DyeDeoxy terminator premix.
3. Spin briefly to collect the reaction mix in the bottom of the tube, and overlay with approx 50 μL of mineral oil.
4. Place the tubes in a thermal cycler (Perkin Elmer Cetus [Warrington, UK] model 480 or 9600), which has been preheated to 96°C.
5. Immediately begin the cycle sequencing program which is as follows:
   a. Rapid thermal ramp to 96°C.
   b. 96°C for 30 s.
   c. Rapid thermal ramp to 50°C.
   d. 50°C for 15 s.
   e. Rapid thermal ramp to 60°C.
   f. 60°C for 4 min.
   g. For a total of 25 cycles.
6. Try to keep the samples in the dark at 4°C until further processing since they are light-sensitive.
7. Remove the excess DyeDeoxy terminators from the completed sequencing reactions.

3.3. Phenol/Chloroform Extraction of Cycle Sequencing Products

This step is essential to remove excess primers and unincorporated nucleotides.

1. To each sample, add 80 μL of sterile dH_2O.
2. Either add 100 μL chloroform to dissolve the oil or remove the oil with a pipet.
3. Add 100 μL phenol:H_2O:chloroform (68:18:14) to the sample and mix well by vortexing.
4. Centrifuge the sample at 12,000g for 1 min. Remove and discard the lower organic phase.
5. Re-extract the aqueous layer, and transfer the aqueous upper layer to a clean tube.
6. Add 15 μL of 2M sodium acetate and 300 μL of 100% ethanol to precipitate the extension products (*see* Note 7).
7. Centrifuge at 12,000g for 15 min at room temperature.
8. Carefully remove the supernatant, and wash the pellet with 70% ethanol. Then briefly dry the pellet under a vacuum (*see* Note 2).

3.4. Preparation of Samples for Loading

1. Add 4 μL deionized formamide: 50 mM EDTA, pH 8.0 (5:1), to each sample tube, and mix well to dissolve the dry pellet.
2. Centrifuge briefly to collect the liquid at the bottom of the tube.
3. Before loading, heat the samples at 90°C for 2–3 min to denature. Then transfer immediately onto ice.
4. Load all of the samples onto the automated DNA sequencer fitted with a 6% polyacrylamide gel using the manufacturer's software.

3.5. Preparation of 6% Polyacrylamide Sequencing Gel (see Note 3)

1. Place 40 g urea, 12 mL 40% acrylamide stock, 20 mL dH_2O, and 1 g mixed-bed ion-exchange resin into a beaker, and stir gently with warming. Continue to stir the solution until all the urea crystals have dissolved (*see* Notes 8 and 9).
2. Filter the acrylamide through a 0.2-μM filter, degas for 5 min, and transfer to a 100-mL cylinder.
3. Add 8 mL of filtered 10X TBE buffer, and adjust the volume to 80 mL with dH_2O.
4. To polymerize the gel, add 400 μL 10% APS (freshly made) and 45 μL TEMED. Gently swirl to avoid adding air bubbles.
5. According to the instructions provided for the automated DNA sequencer, run the sequencing gel and analyze the readouts as indicated in Fig. 1 (*see* Note 10).

4. Notes

1. Both purification steps can also be done by spin columns, such as Centri-Sep or Quick Spin. Although they provide a quicker purification, they are an expensive alternative for those on a tight budget.
2. The dried sequencing pellet can be stored in the dark at 4°C for several days if required. However, once the loading buffer has been added, the samples should be loaded within a few hours.

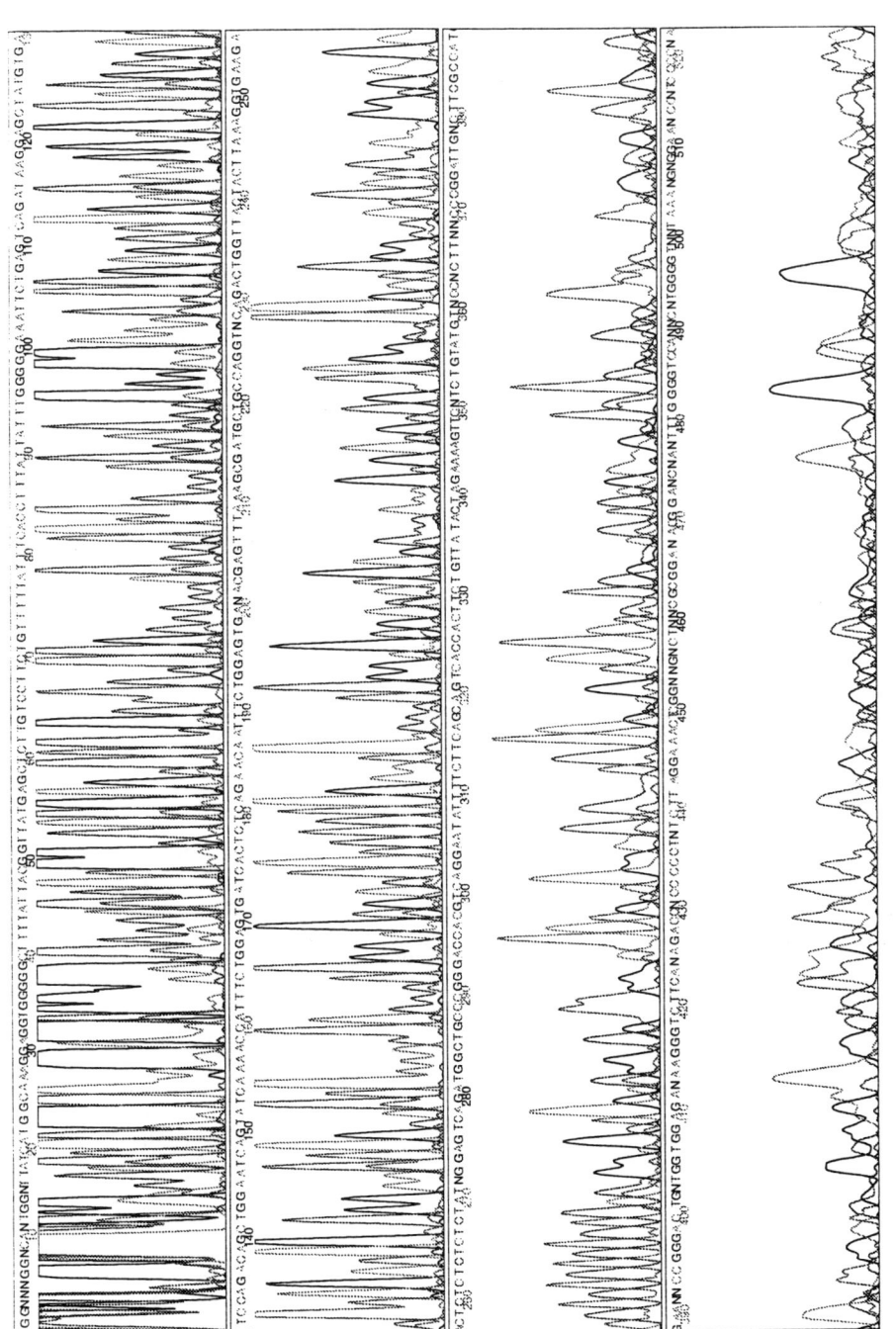

Fig. 1. Analyzed sequence data for a 350-bp PCR fragment amplified from genomic human DNA. The data were obtained using one of the primers used for PCR amplification.

159

3. The sequencing gel must polymerize for at least 1 h before use. A good time to prepare the gel is during the PCR of the cycle sequencing reactions.
4. The PCR primers can be used as the DNA sequencing primers, and should be at least an 18 mer in length. Increasing the length will increase specificity and prevents priming at a secondary site. It will also decrease the chances for nonexact hybridization.
5. The GC content of the primer should be between 50 and 60%.
6. During the phenol/chloroform extraction, if after the first spin two separate layers are not seen, then revortex the samples for 1 min and recentrifuge, after which you should obtain an aqueous and organic phase.
7. After ethanol or isopropanol precipitation, it is very important that the supernatant be carefully aspirated, since the pellets are unstable and might be lost.
8. Although there are various ready-made acrylamide solutions on the market, it is recommended that you make your own solutions, since better resolution will be achieved. However, 40% acrylamide can be purchased ready-made and gives good results.
9. It has been shown that the use of formamide in the polyacrylamide gels can resolve compressions *(13)*.
10. If using the ABI 373A "Stretch" automated DNA sequencers, then it is better to use a 4.75% polyacrylamide gel, and from these machines, it is possible to obtain up to 1 kb of sequence with one run.
11. If the thermal cycler in your laboratory is not a Perkin Elmer Cetus, then it will be necessary to optimize your thermal cycler for the sequencing reactions.
12. The addition of blue dextran to the loading buffer will make gel loading easier.

Acknowledgment

The author would like to thank Applied Biosystems for their constant assistance.

References

1. Orita, M., Suzuki, Y., Sekeiya, T., and Hayashi, K. (1989) Rapid and sensitive detection of point mutations and DNA polymorphisms using the polymerase chain reaction. *Genomics* **5,** 874–879.
2. Kwok, P. Y., Carlson, C., Yager, T. D., Ankener, W., and Nickerson, D. A. (1994) Comparative analysis of human DNA variations by fluorescence based sequencing of PCR products. *Genomics* **23,** 138–144.
3. Martin-Gallardo, A., McCombie, W. R., Gocayne, J. D., Fitzgerald, M. G., Wallace, S., Lee, B. M. B., Lamerdin, J., Trapp, S., Kelly, J. M., Liu, L. I., Dubrick, M., Johnston-Dow, L. A., Kerlavage, A. R., de Jong, P., Carrano, A., Fields, C., and Venter, J. C. (1992) Automated DNA sequencing and analysis of 106 kilobases from human chromosome 19q 13.3. *Nature Genetics* **1,** 34–39.
4. NIH/CEPH Collaborative Mapping Group (1992) A comprehensive genetic linkage map of the human genome. *Science* **258,** 67–86.
5. Mullis, K. B. and Faloona, F. A. (1987) Specific synthesis of DNA in vitro via a polymerase catalysed chain reaction. *Methods Enzymol.* **155,** 335–350.

6. Sanger, F., Nicklen, S., and Coulson, A. R. (1977) DNA sequencing with chain terminating inhibitors. *Proc. Natl. Acad. Sci. USA* **74,** 5463–5467.
7. Maxam, A. M. and Gilbert, W. (1977) A new method for sequencing DNA. *Proc. Natl. Acad. Sci. USA* **74,** 560–564.
8. Carothers, A. M., Urlaub, G., Mucha, J., Grunberger, D., and Chasin, L. A. (1989) Point mutation analysis in a mammalian gene:rapid preparation of total RNA, PCR amplification of cDNA, and *Taq* sequencing by a novel method. *Biotechniques* **7,** 494–499.
9. McBride, L. J., Koepf, S. M., Gibbs, R. A., Nyugen, P., Salser, W., Mayrand, P. E., Hunkapiller, M. W., and Kronick, M. N. (1989) Automated DNA sequencing methods involving polymerase chain reaction. *Clin. Chem.* **35,** 2196–2201.
10. Tracy, T. E. and Mulcahy, L. S. (1991) A simple method for direct automated sequencing of PCR fragments. *Biotechniques* **11(1),** 68–75.
11. Rosenthal, A. and Charnock Jones, D. S. (1992) New protocols for DNA sequencing with dye terminators. *DNA Sequence* **3,** 61–64.
12. Kelley, J. M. (1994) Automated dye terminator DNA sequencing, in *Automated DNA Sequencing and Analysis* (Adams, M. A., Fields, C., and Venter, J. C., eds.), Academic, London, pp. 175–181.
13. Hawkins, T. L. and Sulston, J. E. (1991) The resolution of compressions in automated fluorescent sequencing. *Nucleic Acids Res.* **19(10),** 2784.

21

Solid-Phase Automated Sequencing of PCR-Amplified Genomic DNA

Mary I. Coolbaugh Murphy,
Holly A. Hammond, and C. Thomas Caskey

1. Introduction

The use of the polymerase chain reaction (PCR) allows isolation and examination of genomic DNA sequences using specific oligonucleotide primers to amplify a target region with *Taq* DNA polymerase *(1,2)*. The PCR products can be screened for mutations using molecular techniques, such as allele-specific oligonucleotides (ASO), single-stranded conformation polymorphisms (SSCP), denaturing gel electrophoresis (DDGE), restriction-fragment-length polymorphisms (RFLP), short tandem repeats (STRs), and heteroduplex formation *(3,4)*.

However, these techniques can be uninformative or cost-prohibitive, particularly in the case of new mutation disorders. In circumstances such as these, mutation detection by sequencing may be the most direct, informative, and cost-effective route. We have found screening by methods such as SSCP can help to determine which region of a large gene should be the target of sequencing efforts *(5–7)*.

Direct sequencing of PCR products with automated techniques can be hindered by numerous factors. These include:

1. Using PCR primers for sequencing primers, resulting in low signal or unintelligible sequence reads.
2. Large PCR reaction volumes, which contain adequate sequence template, generally exceed the limiting volumes required for good sequencing reactions.
3. The presence of excess dNTPs, primers, genomic DNA, salts, or *Taq* enzyme from the initial amplification that can interfere with the sequencing chemistry.

From: *Methods in Molecular Biology, Vol. 65: PCR Sequencing Protocols*
Edited by: R. Rapley Humana Press Inc., Totowa, NJ

To resolve these problems, we have chosen to purify PCR products by using the following method:

1. Biotinylate one of the initial PCR primers.
2. Bind the biotinylated PCR product to paramagnetic streptavidin-coated beads. This allows removal of excess primers, dNTPs, enzyme, and salts while concentrating the target DNA into a suitable volume for automated sequencing.
3. Denature and separate the double-stranded PCR product, yielding single-stranded DNAs for Sequenase sequencing *(8–11)* (Fig. 1).

Combining this approach with the use of an automated sequencing device has several benefits. It is possible to automate the amplification setup, the product binding, purification, and sequencing steps utilizing a robot for high through-put. Our approach results in purification of sense and antisense strands of the template for sequencing. Fluorescent sequencing primers eliminate the need for radioactive products *(12–16)*.

An automated sequencing approach incorporates data collection, analysis, and storage into one package. This data package can be further analyzed, compared, displayed, printed, or submitted to a data base. These tasks can be achieved using any of several software packages available commercially. The ease of developing an interface between data and on-line resources is an important factor when building a DNA sequence or mutation data base.

2. Materials

2.1. Genomic DNA Isolation

1. 1X Phosphate-buffered saline (Gibco), 4°C.
2. Lysis buffer: 10 mM Tris-HCl, pH 8.0, 400 mM NaCl, 2 mM Na$_2$ EDTA, filter-sterilized and stored at 4°C.
3. 10% Sodium dodecyl sulfate, filter-sterilized, stored at room temperature.
4. Proteinase K, 100 mg/mL (Boehringer Mannheim Biochemicals), –20°C.
5. Saturated NaCl (>6M), stored at room temperature.
6. Ethanol, 100%, nondenatured (Aaper Chemicals), stored at room temperature.
7. Distilled water, filter-sterilized, stored at room temperature.

2.2. Oligonucleotide Primers

1. Underivatized and 5'-biotinylated primers specific to the genomic target, 25-µM in water, stored in 100 µL aliquots at –70°C.
2. 5'-FAM-labeled nested sequencing primers, 1 µM in water, stored in 200-µL aliquots at –20°C.

2.3. Genomic DNA Amplification

1. PCR-quality water (autoclaved and/or filter-sterilized), stored at room temperature in 1-mL aliquots.

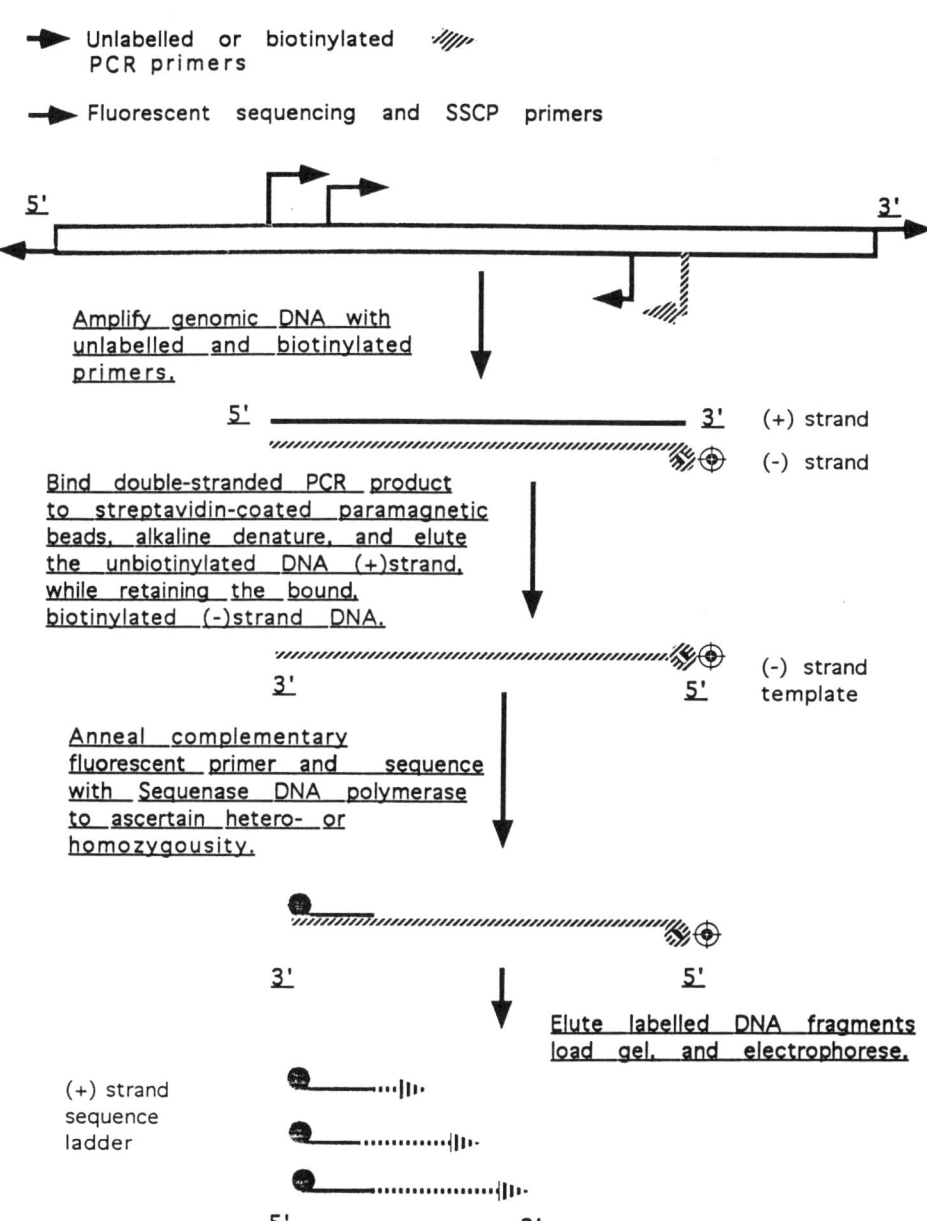

Fig. 1. Schematic flow-chart illustrating the procedure for producing and isolating biotinylated, single-stranded PCR product for genomic sequencing to detect specific mutations.

2. 10X Cetus buffer: 100 m*M* Tris-HCl, pH 8.3, 500 m*M* KCl, 15 m*M* MgCl$_2$, 0.01% w/v gelatin; filter-sterilize and store in 1-mL aliquots at 4°C.
3. 25 m*M* deoxynucleoside triphosphates (dNTPs) mix (dATP + dCTP + dGTP + dTTP) (Pharmacia LKB), stored in 50–µL aliquots at –20°C.
4. AmpliTaq DNA polymerase, 5 U/µL (Perkin-Elmer), stored at –20°C.
5. 50% DMSO in water (Sigma), filter-sterilized, stored in 1-mL aliquots at –20°C.

2.4. Estimating PCR Product Quantity

1. 1% agarose, 1X TBE, 2.5 µg/mL ethidium bromide gel mix, prepared as needed.
2. 10X ALF-TBE: 1*M* Tris-HCl, 0.83*M* boric acid, 10 m*M* disodium EDTA (Bio-Rad, Hercules, CA), filtered and stored at room temperature.
3. Phi-X 174 *Hae*III DNA size markers, 250 µg/µL (Pharmacia), stored at 4°C.
4. Agarose gel loading buffer: 0.25% bromophenol blue, 0.25% xylene cyanol, 30% glycerol, and 1X TBE. Stored in 1-mL aliquots at room temperature.

2.5. Purification of PCR Product

1. 10 mg/mL streptavidin conjugated to paramagnetic beads (DYNAL AS).
2. TTL buffer: 6*M* LiCl, 300 m*M* Tris-HCl, pH 8.0, 0.3% Tween-20, filter-sterilized and stored at room temperature.
3. TT buffer: 10 m*M* Tris-HCl, pH 8.0, 0.1% Tween-20, filter-sterilized and stored at room temperature.
4. 0.2*N* NaOH, prepared fresh.
5. 1X Sequencing TE: 10 m*M* Tris-HCl, pH 7.5, 0.2 m*M* disodium EDTA, filter-sterilized and stored in 1-mL aliquots at room temperature.
6. DYNAL MPC-M and MPC-9600 96-well diurydiun boron magnetic separators.

2.6. Sequencing Gel and Running Buffer

1. 65 mL ALF sequencing gel: 6.7% (19:1) acrylamide-bis mix (Accugel, National Diagnostics), 7.5*M* urea, 1X ALF-TBE buffer: prepared weekly, treated with Amberlite MB-1 (Sigma), filtered, and stored away from light at 4°C.
2. 1X ALF-TBE gel running buffer, 2 L, prepared fresh.
3. 10% Ammonium persulfate, in water, 2 mL, prepared weekly, filtered, and stored at 4°C.
4. TEMED (Bio-Rad), stored at 4°C.

2.7. Sequencing of Single-Stranded PCR Product

1. AutoRead Sequencing Kit (Pharmacia).
2. Automated Laser Fluorescence (ALF) sequencing device (Pharmacia).

2.8. Analysis and Storage of Results

1. 486/66 PC-compatible computer (COMPAQ).
2. ALF Manager v.1.6 (Pharmacia).
3. Windows 3.1 and OS/2 2.1 operating systems (Microsoft and IBM).

3. Methods

3.1. Genomic DNA Preparation

1. Centrifuge tissue-culture cells or leukocyte cell suspension from whole blood at 1000g for 5 min at 10°C in a Beckman J-6B centrifuge.
2. Remove supernatant, wash cells 1× with cold 1X PBS, and centrifuge as in step 1.
3. Remove supernatant, and resuspend cells in 1 mL cold nuclei lysis buffer/T25 flask of cultured cells or/100 μL of leukocytes. Transfer cell mixture to a 1.5-mL Eppendorf microcentrifuge tube. Shake and vortex cells, add 5 μL proteinase K (20 mg/mL) and 80 μL 10% SDS. If solution is viscous and thick, split into two tubes and add lysis buffer to 1 mL. Then add proteinase K and SDS.
4. Mix by vortexing.
5. Incubate at 50°C for 2–4 h.
6. Add 1/3 vol saturated NaCl (approx 400 μL).
7. Centrifuge at 4000g in a microfuge for 20 min at room temperature.
8. Transfer the supernatant to a 15-mL Corning tube, avoiding the precipitated proteins.
9. Add 3 vol of room temperature 100% ethanol (approx 3 mL).
10. Spool out DNA on a glass rod, and transfer to a 1.5-mL tube.
11. Wash DNA twice with a small volume (50–100 μL) of room temperature 70% ethanol.
12. Dry DNA briefly, 2–4 min in a SpeedVac with the heater "on," and resuspend into 500 μL PCR-grade water.
13. Quantitate the DNA by measuring absorbance at 260 and 280 nm. If the 260/280 ratio is <1.8, repeat the isolation process from step 3.
14. Dilute the DNA to 100 ng/μL with 0.1X Sequencing TE.

3.2. Oligonucleotide Preparation (see Note 1)

All oligonucleotide primers were synthesized at the Nucleic Acid Core Laboratory, Baylor College of Medicine, Houston, TX. The primers were produced using phosphoramidite chemistry on an ABI 393 Synthesizer. The biotinylated primers were purified over Nensorb Nucleic Acid columns (DuPont), whereas the FAM-labeled primers were polyacrylamide gel-purified.

3.3. Genomic DNA Amplification (see Note 2)

1. For each target region of DNA to be amplified, prepare a Master Mix of all the reagents, except the DNA (Table 1).
2. In a PCR tube on ice (0.2 mL, Strip-ease, Robbins), place 96 μL of Master Mix, then 4 μL of genomic DNA (200–400 ng), mix, and divide into 4 × 25 μL reactions.
3. Cycle conditions:
   94°C/2:30 1X
   Ramp 0:40
   94°C/0:15
   Ramp 0:40 30X
   55°C/0:15
   Ramp 0:40

Table 1
Reaction Mixes (Total vol = 100 µL)

Reagent	Stock	Final	µL/Reaction
PCR water		q.v. 100 µL	72
Cetus buffer	10X	1X	10
DMSO	50%	7%	14
dNTP	25 m*M*	250 µmol	1
5'-Sense primer	25 µ*M*	25 pmol	1
3'-Antisense primer	25 µ*M*	25 pmol	1
AmpliTaq pol.	5 µ/µL	5 U	1
Volume			100

$72°C/1:00$
$72°C/5:00$ 1X
4°C hold (amplification time is approx 2 h)

4. Turn on thermal cycler (Perkin-Elmer 9600), and allow to preheat. Start program. When the heating lid and block are at 90°C, place tubes in cycler.

3.4. PCR Product Estimation (see Note 3)

1. Mix 5 µL of PCR mixture with 1 µL of agarose gel loading buffer (LB).
2. Mix 2 µL of Phi-X 174 *Hae*III DNA marker with 1 µL LB.
3. Place agarose gel in gel buffer tank, and add 1X TBE buffer just to submerge gel.
4. Load PCR-LB mix into wells of 1% agarose gel in 1X TBE, loading markers into outer wells.
5. Using a squeeze bottle, gently cover the gel with enough 1X TBE to submerge the gel, being careful not to disturb the loaded wells.
6. Electrophorese for 1 h at 60 V (20 V/cm migration desired, not to exceed 80 V without cooling of the minigel apparatus).
7. Rinse gel in water briefly, 2–5 min, and visualize Ethidium bromide stained DNA on a UV light box, using gloves and a UV-blocking face shield.
8. Photograph the gel if desired.
9. Estimate yield of product by comparing relative fluorescence to Phi-X Marker bands. From 500 ng of marker loaded onto the gel, the 330-bp band represents approx 30 ng of DNA. The majority of the products we amplify fall within the 300–400 bp range.

3.5. Solid-Phase PCR Product Purification (see Notes 8 and 9)

1. In a 1.5-mL microcentrifuge tube, wash 40 µL of Dynal streptavidin beads with 100 µL TTL buffer/template to be purified. (The beads are prepared in bulk by increasing the volumes based on the number of templates to be processed.)
2. Apply a Dynal magnet (MPC-M) for 30 s to pull the beads to the side (not the bottom) of the tube, and pipet off the supernatant. Repeat steps 1 and 2.

3. Resuspend the beads in 100 μL of TTL for each template, and mix well.
4. For each template, in one well of the 96-well PCR tray, combine the remaining 95 μL of the biotinylated PCR products (approx 100 ng to 1 μg DNA) with 95 μL of the diluted bead-TTL solution. Mix by pipeting one to two times.
5. Place the PCR tray in a preheated (to 37°C) thermal cycler, and allow the products to bind to the beads for 30 min, mixing with a multichannel pipeter every 10 min.
6. Place the PCR tray on the magnet (MPC-9600), and allow the beads to be pulled to the tube sides for 1 min. Remove the supernatant. Wash the beads twice with 100 μL of TTL, and then remove all supernatant.
7. Resuspend the beads, for each reaction, in 150 μL 0.15*N* NaOH to denature the double-stranded DNA. Incubate at room temperature for 30 min, mixing every 10 min.
8. Remove and save the supernatant for sequencing the complementary strand. (Continue to step 11.)
9. Wash the beads with 100 μL 0.2*N* NaOH. Then neutralize by washing twice with 100 μL of TT buffer, and then wash twice with 100 μL ddH$_2$O.
10. Resuspend the beads in 10 μL of ddH$_2$O, and proceed to sequencing reactions or store the biotinylated templates at 4°C.
11. To neutralize the nonbiotinylated complementary strand, add 10 μL 1.5*N* HCl, mix, and then add 10 μL TE. Precipitate the single-stranded DNA with 0.1 vol 2.5*M* sodium acetate and 2.5 vol cold 100% ethanol. Place on ice for 15 min. Pellet the DNA by centrifugation at 4°C. Remove the supernatant and discard. Wash the pellet in 50 μL 70% ethanol, dry briefly for 2–5 min in a SpeedVac with the heater "on," and resuspend the DNA in 10 μL ddH$_2$O. Proceed to sequencing or store the template at 4°C *(17–19)*.

3.6. ALF Sequencing Gel Preparation

1. Prepare sequencing plates by thoroughly washing with 0.01% Liquinox and a nonscratch scrubbing sponge. Rinse well with ddH$_2$O. Polish dry with Kimwipes. Remove residues with 1 mL isopropanol, and then polish dry. Rinse very well with ddH$_2$O. Polish dry.
2. Assemble plates, 0.3-mm spacers, and light-coupler. Clamp tightly, and place on a sturdy support to raise off bench top (such as two empty styrofoam Corning tube racks). Check to make sure assembly is level.
3. Pour 60 mL of acrylamide mix into a 100-mL squeeze bottle. Add 30 μL TEMED, and 300 μL ammonium persulfate 10%. Swirl gently to mix. Pour gel horizontally by squeezing bottle back and forth along the notched plate end of plate assembly, while rapping knuckles of the other hand along the migrating gel solution front. To remove any bubbles, construct a "fishing hook" by punching a hole on one edge of a retired 0.3-mm spacer. Use this to fish out bubbles. Insert comb, aligning the wells to the etched wells on the ALF plates, and clamp in place. Allow to polymerize for 2 h.
4. After gel is polymerized, clean any residue off light coupler and plate that will be against the light detectors of the ALF, using ddH$_2$O.

5. Attach upper buffer chamber to plates, and place lower buffer chamber in ALF. Place plates in ALF, clamping to secure. Adjust laser. Remove comb. Pour 1 L 1X ALF buffer into each buffer chamber. Rinse wells. Preheat water in the thermoplate to 40°C.

3.7. Sequencing Reactions (see Notes 3, 5, 6, and 7)

1. Preheat two thermoblocks to 37 and 90°C.
2. Remove AutoRead Kit from –20°C and place T7 polymerase on ice. Allow remaining components to thaw at room temperature for approximately 15 min. Briefly centrifuge component tubes, and place on ice. Repeat for fluorescein-labeled primers while keeping shielded from light.
3. Place 10 tubes in a 96-well PCR tray, and place on ice.
4. Adjust volume of magnetic beads to 13 µL with sterile water, for each sample to be sequenced.
5. In a 400-µL microcentrifuge tube, prepare a master mix consisting of 3 µL exon-specific fl-primer (1 pmol/µL)/template and 2 µL annealing buffer/template.
6. Aliquot 5 µL of the master mix to each PCR tube. Add 13 µL template to each tube.
7. Anneal primer to template in a thermal cycler with the following program: 80°C/2:30 → 65°C/6:00 → (ramp 10:00) → 37°C/5:00 → ice.
8. Pipet 2.5 µL of the "A" mix, "C" mix, "G" mix, and "T" mix to each of four respective wells on a microsample tray (Terasaki plate, Nunc), repeating for each template. Keep on ice until ready to preheat these mixes. Before proceeding to the sequencing reactions, place a few drops of mineral oil on the flat surface of the 37°C block, float the microsample tray on the oil, and preheat the d/ddNTP mixes for 1 min (*see* Notes 10 and 11).
9. On ice, add 1 µL extension buffer, 2 µL DMSO, and 2 µL T7 polymerase (3 U) to each template.
10. Mix by pipeting the labeled template reactions. Add 4.5 µL of the annealed primer/template to each of the prewarmed sequencing mixes. Work quickly.
11. Incubate for 5 min at 37°C.
12. Add 3.5–5 µL of Stop solution to each reaction and mix. Use a fresh tip for each addition.
13. Keep microsample tray on ice and shielded from light until ready to load gel, but for no more than 1.5 h.
14. Place a few drops of mineral oil on the flat surface of the 90°C heat block, float microsample tray on the oil and denature the sequencing reactions for 2 min and 30 s. Then place quickly on ice, shielded from light.
15. Rinse wells of gel, and load 5 µL of a "dummy" reaction, consisting of a sequencing reaction using water instead of DNA, into the outermost well on each side of the gel. Rinse four wells, and load 5–7 µL/well for the first set of reactions. Repeat until all 10 sets have been loaded onto the gel.
16. Lower the upper buffer chamber cover, connect the electrode leads securely, check the clamps, water connections, and laser focus. Close the ALF covers.
17. Start run *(15,20–22)*.

3.8. Sequencing Gel Run Conditions

Depending on length of product, run time equals 4–6.5 h for 320–520 bp of sequence. The running conditions are: V = 1200, mA = 36, W = 42, temperature = 50°C, laser = 4 mW.

3.9. Sequence Analysis (see Note 14)

1. The data are collected and analyzed using the OS/2-based ALF Manager v.2.6.
2. After the run, the data are displayed as a four-color chromatogram on the monitor screen in the processed mode. Heterozygous variations from wild-type sequence are specifically identified and highlighted by the automated ALF-Manager program, which uses the IUPAC codes for nucleotide mixtures. The ALF-Manager-detected mutations can be rapidly visualized either by scanning normal samples on the screen in parallel or by looking for variations in the IUPAC code from the usual "A," "C," "G," or "T." It is possible to do both types of scans simultaneously. One can easily scan three samples and one normal on the same screen at one time (*see* Figs. 2 and 3).
3. The processed data can be set to print out automatically immediately after the run. After scanning the print-outs, specific areas of individual sequences can be chosen to print alongside kindred's or control sequences for comparison as described in step 2. The time spent editing is greatly reduced using this approach.
4. The processed data are automatically exported, in GCG format, to a UNIX-based network/storage disk. The ALF work station is cross-mounted onto this networked system.
5. After exiting the ALF Manager software and starting the network software, the data are sent to a modified resident UNIX script of "Mailfasta" *(23)*, which subjects each sequence to BLAST and FASTA data-base searches. *(24)*. These analyses compare the results to the published gene sequence, and can confirm the presence or absence of sequence variations. This process also identifies, in an automated manner, homozygous sequence variations, such as homozygous polymorphisms and mutations.

4. Notes

1. Biotinylated primers are sensitive to multiple freeze–thaw cycles. To preserve the primers and to prevent PCR contamination, they were aliquoted and stored at −70°C until needed.
2. Primer design: We used Primer 2.0 (Eric S. Lander, MIT) and Oligo v. 4.0 (National Biosciences) software to design the primers, using the parameters of 0 mismatches, <3 bp self-annealing at the 3'-end, <3 bp 3'–3' annealing of the PCR primers to each other, no GC clamps, 20–25 bp as the optimum primer length, 50% GC content, and 55–68°C melting temperatures. We sought to design primers that were kinetically more stable at the 3'-terminus. Optimal placement of the PCR primers was with the 5'-end approx 50 bp from the intron–exon junction, then nesting the labeled sequencing primer at least 10 bp 3' to the 5'-end of the

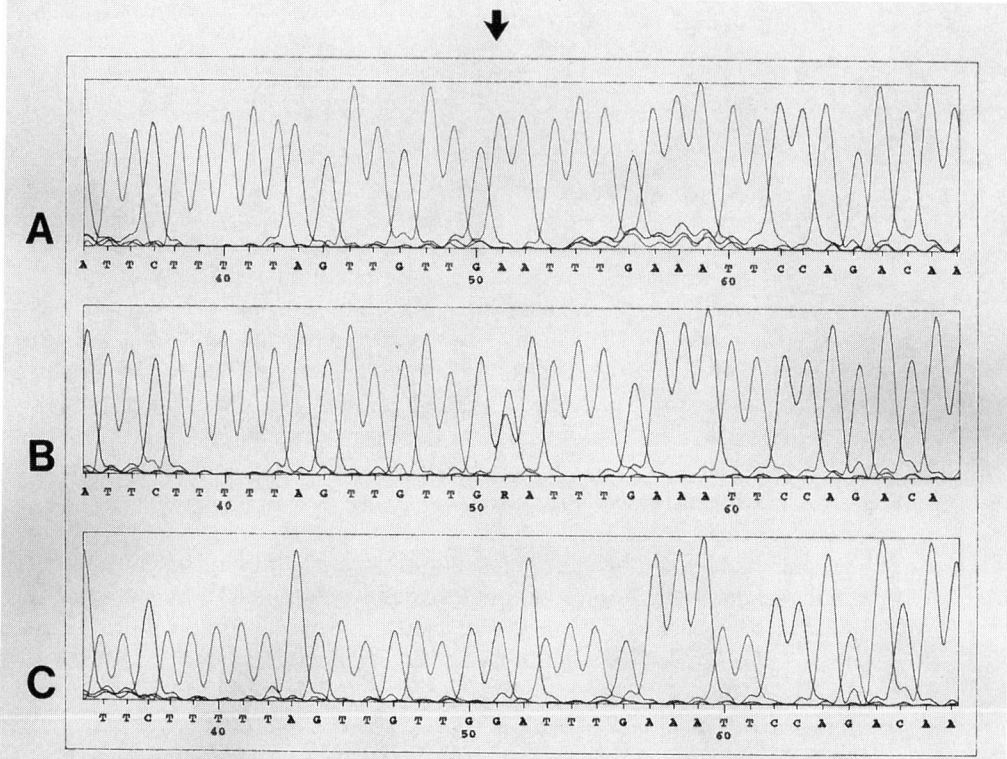

Fig. 2. Trace scan demonstrating the resolution of a mutation in HPRT exon 8 using Sequenase based sequencing chemistry and fluorescently labeled primers on a Pharmacia ALF DNA sequencer. The top chromatogram **(A)** contains sequencing results from an affected male, with a G-to-A transition at base-call number 51. The middle chromatogram **(B)** contains results from his mother, and reveals a G/A heterozygous site at base-call number 51. The bottom chromatogram **(C)** contains sequence from an unaffected female and displays the wild-type G at base-call number 51.

PCR primer. Reads generally started within 5 bp of the 3'-end of the sequencing primer, so this placement allowed analysis of splice junctions *(25)*.

3. The use of DMSO with both T7 and *Taq* polymerases can help resolve GC-rich or repeated areas by lowering the melting temperature of both DNA–primer hybridization complexes and strong secondary structures associated with high GC content. This allows even T7 polymerase, with its lower extension temperature, to produce termination products *(26)*.

4. Agarose gel check: This is done to ensure that only one amplification product is present and that most of the primers have been incorporated. Unrelated fragments with priming sites can result in an ambiguous sequence read *(2,27)*.

5. T7 DNA polymerase may be preferable over *Taq* DNA polymerase because it offers even peak heights and low background, which allows reliable heterozy-

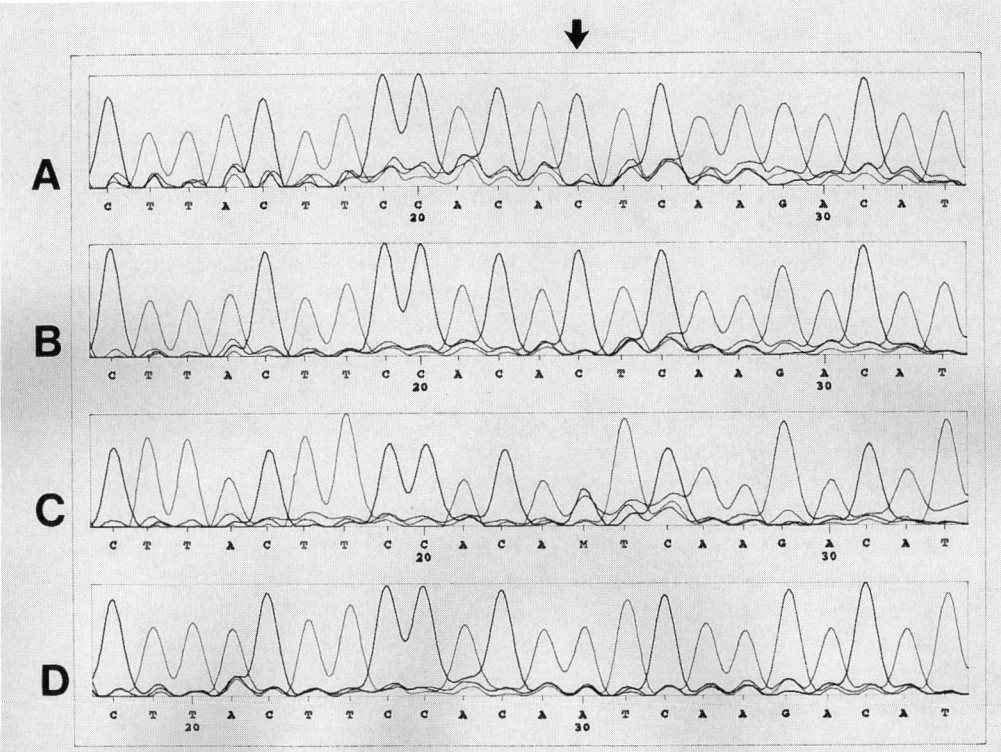

Fig. 3. Trace scans demonstrating the resolution of a kindred's mutation in HPRT exon 5. The Sequenase based sequencing chemistry used fluorescently labeled primers on a Pharmacia ALF DNA sequencer. The top chromatogram **(A)** contains sequencing results from an affected male, with an A-to-C transversion at base-call number 24. The **(B)** chromatogram contains prenatal results from a male sibling, displaying the same mutation. The **(C)** chromatogram shows prenatal sequencing results from their female sibling, a Lesch-Nyhan carrier, as demonstrated by her A/C heterozygous site at base-call number 24. The bottom chromatogram contains sequence from an unaffected, unrelated female who displays the wild-type A at base-call number 30.

gote detection and resolution. An advantage of T7 polymerase is that *Taq*'s higher reaction temperature could cause unstable primer-template complexes, resulting in lower signal intensity, whereas the T7 sequencing annealing procedure allows the formation of more stable primer–template hybrids, resulting in stronger signal intensity *(28)*. *Taq* polymerase allows the use of less starting material, but can produce a higher background and uneven peak heights, which can mask heterozygote detection *(29)*.

6. The addition of deaza-deoxyguanosine to the sequencing reactions can help resolve compressions in the sequencing reactions using either T7 or *Taq* polymerases. However, sequencing with deaza-deoxyguanosine of PCR products that

also have incorporated deaza-deoxyguanosine has, in our experience, less than optimal results *(30)*.

7. It has been observed that difficulty in sequencing double-stranded linear DNA, i.e., PCR products, may be owing to large amounts of complementary strands competing with the sequencing primers to associate with the template *(28)*.

8. To bind the needed 3–5 pmol of single-stranded template requires at least 20 µL magnetic beads (at 10 mg/mL), plus a sufficient excess to bind any unincorporated biotinylated PCR primer. Unincorporated biotinylated primer, compared to the biotinylated PCR product, will bind preferentially to the streptavidin-coated magnetic beads, possibly because of its smaller base-pair size. Using an excess of the unmodified primer in the PCR reaction can be another approach to resolve this problem *(3,19)*.

9. Once the PCR product is captured onto a solid support, it is stable for at least 1 wk, and up to 1 mo in our experience. This allows for the preparation and purification of large numbers of templates in advance of the sequencing reactions *(19)*.

10. Exon-specific end-labeled fluorescent primers were used for both SSCP analysis and sequencing of the genomic DNA. Prescreening by fluorescent SSCP allowed sequencing of mutation containing exons in a timely manner *(7)*.

11. Genomic DNA sequencing with fluorescent primers worked optimally in our hands using 3–5 pmol of labeled primer and 3–5 pmol of single-stranded template *(31)*. An estimate of PCR product amounts for varying DNA fragment sizes that work well on the ALF sequencer follows:

    For a 250-bp DNA fragment, use 450 ng PCR product.
    For a 300-bp DNA fragment, use 540 ng PCR product.
    For a 400-bp DNA fragment, use 720 ng PCR product.
    For a 500-bp DNA fragment, use 810 ng PCR product.

12. AutoRead parameters: Optimal template amounts are maximally 1 µg; however, we have obtained good reads from 100 ng of a 250-bp template.

13. Magnetic beads can be loaded on a sequencing gel without termination product migration being distorted *(32)*.

14. The ALF-Manager software's base-calling reliably identifies DNA base mixtures that occur in genetic heterozygotes using the IUPAC codes for nucleotide combinations. The problem of mutation/polymorphism calling has relied heavily on visual editing. Our approach, a combination of the ALF-Manager software and automated data-base searches, addresses the issue of detecting both heterozygous and homozygous variations from wild-type sequence in a manner that does not require extensive visual editing by the user(s). The final results of the sequence analyses produce a print-out of the variations, which are then viewed for accuracy.

References

1. Saiki, R. K., Bugawan, T. L., Horn, G. T., Mullis, K. B., and Erlich, H. A. (1986) Analysis of enzymatically amplified β-globin and HLA-DQα DNA with allele-specific oligonucleotide probes. *Nature* **324,** 163–166.

2. Mullis, K. B., Ferré, F., and Gibbs, R. A. (eds.) (1994) *PCR: The Polymerase Chain Reaction.* Birkhauser, Boston, pp. 395–405.

3. Andersson, B. and Gibbs, R. A. (1994) PCR and DNA sequencing, in *PCR: The Polymerase Chain Reaction* (Mullis, K. B., Ferré, F., and Gibbs, R. A., eds.), Birkhauser, Boston, pp. 201–213.

4. Rossiter, B. J. F. and Caskey, C. T. (1994) Clinical applications of the polymerase chain reaction, in *PCR: The Polymerase Chain Reaction* (Mullis, K. B., Ferré, F., and Gibbs, R. A., eds.), Birkhauser, Boston, pp. 395–405.

5. Gibbs, R. A., Nguyen, P.-N., Edwards, A. O., Civitello, A., and Caskey, C. T. (1990) Multiplex DNA deletion detection and exon sequencing of the hypoxanthine phosphoribosyltransferase gene in Lesch-Nyhan families. *Genomics* **7,** 235–244.

6. Makino, R., Yazyu, H., Kishimoto, Y., Sekiya, T., and Hayashi, K. (1992) F-SSCP: Fluorescence-based polymerase chain reaction-single-strand conformation polymorphism (PCR-SSCP) analysis. *PCR Methods and Applications* **2,** 10–13.

7. Hammond, H. A., Albright, D. B., Coolbaugh Murphy, M., and Caskey, C. T. (1995) Mutation Scanning at the Lesch-Nyhan Locus, HPRT, in preparation.

8. Hultman, T., Stahl, S., Hornes, E., and Uhlen, M. (1989) Direct solid phase sequencing of genomic and plasmid DNA using magnetic bead and solid support. *Nucleic Acids Res.* **17,** 4937–4946.

9. Bevan, I. S., Rapley, R., and Walker, M. R. (1992) Sequencing of PCR amplified DNA. *PCR Methods and Applications* **1(4),** 222–228.

10. Uhlen, M., Hultman, T., Wahlberg, J., et al. (1992) Semi-automated solid-phase DNA sequencing. *TibTech* **10,** 52–55.

11. Lagerkvist, A., Stewart, J., Lagerstrom, M., and Landegren, U. (1994) Manifold sequencing: efficient processing of large sets of sequencing reactions. *Proc. Natl. Acad. Sci. USA* **91,** 2245–2249.

12. Smith, L. M., Sanders, J. Z., Kaiser, R. J., et al. (1986) Fluorescent detection in automated DNA sequence analysis. *Nature* **321,** 674–679.

13. Ansorge, W., Sproat, B. S., Stegemann, J., Schwager, C., and Zenke, M. (1987) Automated DNA sequencing: ultrasensitive detection of fluorescent bands during electrophoresis. *Nucleic Acids Res.* **15,** 4593–4602.

14. Voss, H., Schwager, C., Wirkner, U., et al. (1989) Direct genomic fluorescent on-line sequencing and analysis using in vitro amplification of DNA. *Nucleic Acids Res.* **17(7),** 2517–2527.

15. Ansorge, W., Zimmermann, J., Erfle, H., et al. (1993) Sequencing reactions for ALF (EMBL) automated DNA sequencer, in *DNA Sequencing Protocols* (Griffin, H. G. and Griffin, A. M., eds.), Humana, Totowa, NJ, pp. 317-356.

16. Rolfs, A. and Weber, I. (1994) Fully-Automated, Non-radioactive Solid-Phase Sequencing of Genomic DNA obtained from PCR. *BioTechniques* **17(4),** 782–787.

17. Cha, R. and Thilly, W. (1993) Specificity, Efficiency, and Fidelity of PCR. *PCR Methods and Applications* **3,** S18–S29.

18. AppliedBiosystems (1992) User Bulletin no. 21: Magnetic Beads (Dynabeads) used as solid support in purification and cycle sequencing of PCR products.

19. Dynal (1994) Technical Tips Update 1.1, Dynal, AS.
20. Sanger, F., Nicklen, S., and Coulson, A. R. (1977) DNA sequencing with chain-terminating inhibitors. *Proc. Natl. Acad. Sci. USA* **74,** 5463–5467.
21. Zimmerman, J., Voss, H., Schwager, C., Stegemann, J., and Ansorge, W. (1988) Automated Sanger dideoxy sequencing reaction protocol. *FEBS Lett.* **233,** 432–436.
22. Pharmacia (1994) AutoRead Sequencing Kit, Instruction Booklet. Uppsala, Sweden.
23. de Boer, T. (1993) MailFasta, Amsterdam.
24. Altschul, S., Gish, W., Miller, W., Myers, E., and Lipman, D. (1990) Basic local alignment search tool. *J. Mol. Biol.* **215,** 403–410.
25. Chamberlain, J. S. and Chamberlain, J. R. (1994) Optimization of multiplex PCRs, in *PCR: The Polymerase Chain Reaction* (Mullis, K. B., Ferré, F., and Gibbs, R. A., eds.), Birkhauser, Boston, pp. 38–46.
26. Winship, P. R. (1989) An improved method for directly sequencing PCR amplified material using dimethyl sulfoxide. *Nucleic Acids Research* **17(3),** 1266.
27. Maniatis, T., Fritsch, E. F., and Sambrook, J. (1989) *Molecular Cloning: A Laboratory Manual.* Cold Spring Harbor Laboratory Press, Cold Spring Harbor, NY.
28. Kusakawa, N., Uemori, T., Asada, K., and Kato, I. (1992) Rapid and reliable protocol for direct sequencing of material amplified by the polymerase chain reaction, in *The PCR Technique: DNA Sequencing* (Ellingboe, J. and Gyllensten, U. B., eds.), Eaton, Natick, MA, pp. 17–26.
29. Saiki, R. K., Gelfand, D. H., Stoffel, S., Scharf, S. J., Higuchi, R., Horn, G. T., Mullis, K. B., and Erlich, H. A. (1988) Primer-directed enzymatic amplification of DNA with a thermostable DNA polymerase. *Science* **239,** 487–494.
30. McConologue, L., Brow, M. A. D., and Innis, M. A. (1988) Structure-independent amplification of PCR using 7-deaza-2'-deoxyguanosine. *Nucleic Acids Res.* **16,** 9869.
31. Hyder, S. M., Hu, C., Needleman, D. S., Sonoda, Y., Wang, X.-Y., and Baker, V. V. (1994) Improved accuracy in direct automated DNA sequencing of small PCR products by optimizing the template concentration. *BioTechniques* **17(3),** 478–482.
32. Gyllensten, U. B., Allen, M., and Josefsson, A. (1992) Sequencing of in vitro amplified DNA, in *The PCR Technique: DNA Sequencing* (Ellingboe, J. and Gyllensten, U. B., eds.), Eaton, Natick, MA, pp. 1–15.

22

Cloning PCR Products for Sequencing in M13 Vectors

David Walsh

1. Introduction

Although numerous methods are now available for direct sequencing of PCR products, cloning of amplified DNA for sequencing in M13 vectors remains an attractive approach because of the high quality of sequence information generated from single-stranded bacteriophage DNA templates.

Cloning of PCR products is in theory straightforward, but in practice is often problematical, as widely reported *(1–4)*. Difficulties are generally ascribed to modifications of the DNA termini by *Taq* DNA polymerase. After completion of thermal cycling, the enzyme may remain associated with DNA ends and thus interfere with subsequent ligations, unless specific steps are included for its removal or inactivation. Carryover of *Taq* DNA polymerase and residual dNTPs into restriction digests can also result in end filling of 5'-overhangs *(5)*, severely reducing the efficiency of cohesive-ended cloning strategies utilizing restriction sites within PCR primers. Removal of *Taq* DNA polymerase by proteinase K digestion *(6)* or repeated phenol/chloroform extractions *(5)* circumvents these problems.

Furthermore, the terminal transferase activity of *Taq* DNA polymerase catalyzes the nontemplate-directed addition of a single nucleotide, almost invariably deoxyadenosine (dA), to the 3'-ends of amplified DNA molecules *(7)*. The resulting "ragged ends" must be removed if blunt-end ligation to *Sma*I-cut vector is required. This is best achieved by utilizing the strong 3'–5' exonuclease activity of T4 DNA polymerase, which in the presence of low concentrations of dNTPs removes 3'-overhangs from double-stranded DNA. Cloning of amplified DNA into linearized 5'-dephosphorylated vector also necessitates the presence of 5'-phosphate groups on the PCR products, achieved by kinasing either the primers before amplification or the PCR product itself.

From: *Methods in Molecular Biology, Vol. 65: PCR Sequencing Protocols*
Edited by: R. Rapley Humana Press Inc., Totowa, NJ

Conveniently, 3'-dA removal by T4 DNA polymerase and 5'-phosphorylation by T4 polynucleotide kinase can be performed simultaneously *(8)*.

Difficulties in cloning PCR products as blunt-ended molecules may be avoided by incorporating restriction sites into the PCR primers and cloning products more efficiently as cohesive-ended molecules. The major problem encountered here is the failure of some restriction endonucleases to cleave toward the ends of DNA fragments. The presence of 4 bp 5' to the recognition sequence is sufficient for efficient cutting by most, but not all M13 polylinker enzymes (>75% digestion in 2 h by *Eco*RI, *Kpn*I, *Ava*I, *Xma*I, *Pst*I, *Bam*HI, *Sac*I, and *Xba*I, but < 10% digestion in 2 hours by *Acc*I, *Sph*I, *Sal*I, and *Hin*dIII) *(9,10)*. Cutting efficiency may be improved by use of primers with longer regions 5' to the recognition site, but since enzymes, such as *Xho*I, require 20 bp 5' to the site *(11)*, this solution is not recommended. Rather, inefficiently cut terminal sites should be converted to internal sites by concatamerizing the PCR product with T4 DNA ligase in the presence of PEG *(10,12,13)*. Internalized restriction sites are cut with greatly improved efficiency, facilitating a corresponding increase in cloning efficiency of amplified DNA.

In addition to the required product, the completed PCR reaction mixture contains residual dNTPs and primers, and often artifactual short amplification products and primer-dimers. These unwanted species may be present in molar excess and, if carried forward into DNA-modifying reactions, are likely to result in reduced cloning efficiency of the required product. Thus, it is beneficial to purify the required PCR product prior to modification, either by means of a spin filtration device, such as Microcon™, or by glass bead isolation from agarose using Geneclean™.

This chapter describes the manipulations required for the efficient blunt- and cohesive-ended cloning of PCR products into M13 vectors. Chapter 23 contains protocols for the subsequent sequencing of fragments cloned into M13.

2. Materials

1. Phenol/chloroform: Mix equal volumes of Tris-buffered phenol, pH > 7.5 (nucleic-acid-grade) and chloroform (AR-grade). Store at 4°C in a dark glass bottle.
2. Chloroform/isoamyl alcohol (24:1): Store at 4°C.
3. Spin filtration device, such as Amicon Microcon™ (Beverly, MA) or Promega Wizard™ PCR purification unit (Madison, WI).
4. Variable-speed microcentrifuge.
5. TE buffer: 10 mM Tris-HCl, 1 mM EDTA, pH 8.0.
6. T4 DNA polymerase.
7. 10X T4 DNA polymerase buffer: 500 mM NaCl, 100 mM, Tris-HCl, pH 7.9, 100 mM MgCl$_2$, 10 mM DTT, 500 μg/mL BSA.
8. T4 polynucleotide kinase.

9. 10X polynucleotide kinase (PNK) forward reaction buffer: 500 mM Tris-HCl, pH 7.5, 100 mM MgCl$_2$, 50 mM DTT, 500 µg/mL BSA.
10. T4 DNA ligase.
11. 10X T4 DNA ligase buffer: 500 mM Tris-HCl, pH 7.8, 100 mM MgCl$_2$, 100 mM DTT, 250 µg/mL BSA.
12. Calf intestinal alkaline phosphatase (CIP).
13. 10X CIP buffer: 200 mM Tris-HCl, pH 8.0, 10 mM MgCl$_2$, 10 mM ZnCl$_2$.
14. Restriction enzymes at 10 U/µL and 10X reaction buffers.
15. 0.5M EDTA, pH 8.0.
16. 3M sodium acetate, pH 5.3.
17. Absolute ethanol, stored at –20°C.
18. Ultrapure stock of all four dNTPs: Make 20 mM stock in sterile water and store at –20°C. Avoid freeze/thaw.
19. 10 mM ATP: Store in aliquots at –20°C. Avoid freeze/thaw.
20. 60% (w/v) PEG 8000 in water, filter-sterilized. Store at room temperature away from direct sunlight.
21. Ultrapure low-melting-point agarose.
22. Glass bead DNA isolation kit, such as Geneclean (Bio101, Madison, WI).
23. M13 RF DNA purchased from supplier or prepared in-house by CsCl density gradient centrifugation.
24. Isopropyl-β-thiogalactopyranoside (IPTG): 20 mg/mL stock in sterile water, stored at –20°C.
25. 5-Bromo-4-chloro-3-indolyl-βD-galactoside (X-gal): 20 mg/mL stock in dimethyl formamide, stored at –20°C in glass vial.
26. *Escherichia coli* strain containing the F plasmid (e.g., JM101, JM107) and maintained on M9 minimal agar.

3. Methods

3.1. Purification of PCR Products
from Taq *DNA Polymerase, Primers, and dNTPs (see Notes 1 and 2)*

1. Extract the completed PCR reaction with an equal volume of chloroform. Spin in a microfuge at 13,000g for 2 min to separate the aqueous and organic phases.
2. Extract the upper aqueous phase twice with an equal volume of phenol/chloroform and once with an equal volume of chloroform/isoamyl alcohol.
3. Transfer the upper aqueous layer (up to 100 µL) into a Microcon unit housed in a 1.5-mL Eppendorf tube, and add 400 µL TE buffer. Spin in a microfuge at 500g for 15 min (Microcon-100) or at 14,000g for 6 min (Microcon-50). Add a further 400 µL TE to the sample reservoir, and spin as before. Volume retained will now be 50–100 µL. If required (PCR product present in low yield), concentration down to a volume of 10–20 µL is achieved by a further spin cycle. Each cycle reduces the concentration of salts, PCR primers, and dNTPs by approx 95%.
4. Invert the unit in a fresh tube, and spin at 500g for 2 min to recover purified PCR product.

5. Check recovery by agarose gel electrophoresis of an aliquot of the concentrated product.

If amplified DNA is to be cloned by cohesive-end ligation via restriction sites incorporated into PCR primers, and these sites are known to cut efficiently, the purified product can now be digested with restriction endonucleases without further processing. PCR products to be cloned by blunt-end ligation or via digestion with restriction enzymes that cut inefficiently at DNA termini should be processed as follows.

3.2. Simultaneous End Repair and Phosphorylation of PCR Products (see Notes 3 and 4)

1. Set up a reaction containing: 100 ng to 1 µg purified PCR product, 3 µL 10X T4 DNA polymerase buffer, 100 µ*M* each dNTP, 1 m*M* ATP, 0.5 U T4 DNA polymerase, 5 U T4 polynucleotide kinase, and H$_2$O to 30 µL. Incubate at 25°C for 20 min.
2. Stop the reaction by incubating at 75°C for 10 min in the presence of 5 m*M* EDTA, pH 8.0.
3. Increase the volume to 100 µL with H$_2$O, and perform one extraction with phenol/chloroform and one with chloroform/isoamyl alcohol.
4. Remove the aqueous phase to a fresh tube, and add 0.1 vol 3*M* sodium acetate, pH 5.3, and 2.5 vol cold absolute ethanol. Mix well and store at –20°C for 1 h or at –70°C for 20 min. Precipitate DNA by centrifugation at 13,000*g* for 10 min in a microfuge.
5. Remove the supernatant carefully, and add 0.5 mL cold 70% ethanol to the pellet. Spin again at 13,000*g* for 2 min. Discard the supernatant as before, vacuum dry the pellet (2–5 min), and finally dissolve DNA in 10 µL TE.

PCR products are now flush-ended and phosphorylated at their 5'-ends, ready for direct cloning into *Sma*I-cut, dephosphorylated M13 vector or for concatamerization, as required.

3.3. Concatamerization/Digestion of PCR Products (see Notes 5–8)

1. To 10 µL PCR product add: 2 µL 10X T4 DNA ligase buffer, 7 µL 60% PEG 8000 (20% final), and 2 U T4 DNA ligase. Incubate at room temperature overnight.
2. Increase the volume to 100 µL with water, and perform one extraction with phenol/chloroform. Avoiding the white PEG precipitate at the interface, remove 10 µL of the aqueous phase to check extent of concatamerization by electrophoresis through 0.8% agarose. Run out alongside an aliquot of the original PCR product and DNA size markers (*see* Note 8).
3. If concatamerization is judged to be successful (PCR product present as trimers and larger species), extract the remaining 90 µL once with chloroform/isoamyl alcohol and precipitate with sodium acetate/ethanol as above. Dissolve in 20 µL TE.
4. Add 10 U of appropriate restriction enzyme(s), 3 µL 10X reaction buffer, and H$_2$O to 30 µL. Incubate at the required temperature for 1 h.

If cutting with two enzymes that have different salt requirements, digest with the low-salt enzyme first, heat-inactivate at 65°C for 20 min (or phenol-extract heat-stable enzymes), then adjust salt concentration with 1M NaCl, and add 10 U of the second enzyme. Where two enzymes require completely different buffers, phenol/chloroform-extract and sodium acetate/ethanol-precipitate the DNA in between each digest.

5. Check to ensure the digest now contains only monomer-size PCR product by agarose gel electrophoresis. If larger species remain, indicating incomplete digestion of concatamers, add more enzyme and continue digestion.
6. When digestion is complete, electrophorese the digest mixture through 0.8% low-melting-point agarose, and recover the PCR product by glass bead isolation using Geneclean.

3.4. Preparation of M13 Vector DNA for Ligation

3.4.1. Digestion with Restriction Endonucleases

Where digestion of the vector with two enzymes is required, the ability of each enzyme to cleave toward the end of linear DNA molecules should be considered to determine the preferred order of sequential addition (*see* Note 9) (New England Biolabs catalog, Reference Appendix). Enzymes that cut less efficiently toward DNA termini should be used first. In this situation where directional cloning is required, digest M13 derivatives containing the polylinker in both orientations, e.g., M13mp18 and M13mp19.

1. Set up the following digest: 1 µg M13 RF DNA, 5 U restriction enzyme, 2 µL 10X reaction buffer, and H$_2$O to 20 µL.
2. Incubate for 1 h at the appropriate temperature (25°C for *Sma*I and 37°C for all other polylinker enzymes). Remove 1 µL to analyze extent of digestion by electrophoresis through 0.8% agarose. If digestion is not complete, add more enzyme and continue incubation.
3. When complete, extract the digest once with phenol/chloroform, and precipitate DNA with sodium acetate/ethanol as in Section 3.2., steps 4–6. Dissolve the DNA pellet in 10 µL TE, and digest with a second enzyme if required.

M13 DNA cut with a single enzyme should now be treated with calf intestinal alkaline phosphatase to reduce recircularization during ligation.

3.4.2. Dephosphorylation of Vector DNA

1. To 1 µg linearized M13 DNA in 10 µL TE add: CIP 0.05 U for 5'-overhangs, 0.5 U for 3'-overhangs or blunt ends, 5 µL 10X CIP buffer, and H$_2$O to 50 µL total volume
2. Incubate at 37°C for 60 min.
3. Inactivate CIP by heating the reaction to 75°C for 10 min in the presence of 5 mM EDTA, pH 8.0.
4. Extract the reaction once with phenol/chloroform, and recover DNA by sodium acetate/ethanol precipitation as in Section 3.2., steps 4–6. Dissolve the DNA pellet in 20 µL TE.

5. Check recovery of M13 vector, and insert DNA by electrophoresing an aliquot of each through 0.8% agarose.

3.5. Ligation of PCR Products into M13 Vectors (see Notes 10 and 11)

1. Set up the ligation reaction, add components in the following order: 50 ng M13 vector DNA, 1 μL 10 mM ATP (1 mM final), 1 μL 10X ligation buffer, 1–4 μL DNA insert (three- to fivefold molar excess), 5 U T4 DNA ligase for blunt termini, 1 U T4 DNA ligase for cohesive termini, and H$_2$O to 10 μL total volume.
2. Set up a negative control ligation in which an equal volume of water is substituted for the PCR DNA insert, and a positive control ligation containing an appropriate blunt- or cohesive-ended restriction fragment, preferably of a similar size to the PCR product.
3. Incubate overnight at 14°C for cohesive-end ligations or at room temperature for blunt-end ligations.
4. Transform 2.5–5 μL of the ligation reaction into *E. coli* competent cells, and plate in a soft agar overlay containing 0.33 mM IPTG and 0.03% X-gal. Identify recombinant phage clones by blue/white selection (*see* Note 11).

4. Notes

1. Use of a spin filtration unit to purify the PCR product from unincorporated nucleotides and primers is only recommended if the PCR reaction mixture does not contain unwanted species larger than the nucleotide cutoff value of the unit. For the Microcon-100, this corresponds to 300 bases (single-stranded) or 125 bp (double-stranded). If larger unwanted products are present, the target product should be purified by electrophoresis through low-melting-point agarose followed by adsorption to glass beads.
2 If a spin filtration device is not available, PCR products can be partially purified from residual primers and dNTPs by precipitation with sodium acetate/ethanol, followed by a 70% ethanol wash. Better removal of such reactants can be achieved by adjusting to 2M ammonium acetate and adding 2 vol of ethanol, although it should be noted that ammonium ions are a strong inhibitor of T4 DNA polymerase and must be thoroughly removed by extensive washing in 70% ethanol before the end-repair step.
3. Occasionally, PCR products end-repaired and kinased as described may fail to clone as blunt-ended molecules, most probably because of the persistence of *Taq* DNA polymerase bound at DNA termini. In this situation, residual enzyme can be removed by adding to the sample 50 μg/mL proteinase K in 10 mM Tris-HCl, pH 7.8, 5 mM EDTA, 0.5% (v/v) SDS, and incubating at 37°C for 30 min. Extract with phenol/chloroform, and precipitate PCR products with sodium acetate/ethanol.
4. It is important not to exceed the recommended amount of T4 DNA polymerase enzyme or the incubation time of 20–30 min for the end-repair reaction, since both may result in excessive exonuclease activity and nonblunt "nibbled ends." T4 DNA polymerase also has excessive exonuclease activity at higher temperatures (37°C).

5. For palindromic restriction enzyme sites, concatamerization of PCR products containing terminal half-sites reconstitutes the site. For example, end-to-end ligation of DNA molecules with terminal sequences GGA-3' and 5'-TCC reconstitutes the *BamH*I recognition sequence. This allows the use of shorter PCR primers containing fewer extraneous nucleotides that do not hybridize to the target sequence.

6. Intermolecular joining of PCR products is stimulated by macromolecular exclusion molecules, such as PEG 8000, with maximal stimulation occurring in the range 15–25% (w/v).

7. Concatamerization and digestion of PCR products containing nucleotides 5' to the restriction site generates a small cohesive-ended fragment that can coprecipitate with the full-length product and give rise to false positives on ligation into M13. Purification of the required product by glass bead isolation from low-melting-point agarose prior to cloning is therefore recommended. Eighty to 90% of clear plaques examined should then be found to contain the required insert. This figure falls to 10–20% if purification is not performed.

8. To allow analysis of concatamerized PCR products by gel electrophoresis, PEG must first be removed by extracting with phenol/chloroform, since the presence of the polymer prevents entry of DNA into agarose.

9. When digesting M13 RF DNA with two enzymes that cut at closely spaced sites in the polylinker, it is preferable to perform the digests sequentially, rather than simultaneously, even when both enzymes are active in the same buffer, in order to maximize the cutting efficiency of each enzyme.

10. The number of clear plaques that can be expected depends on the strategy being followed. Cloning via direct cutting of efficiently recognized terminal restriction sites should yield 50–100 clear plaques per transformation. Rather fewer, typically 10–20, are produced by the concatamerization/digestion and blunt-end approaches.

11. M13 clones containing the required insert can quickly be identified by PCR using primers employed for the original amplification. Use sterile toothpicks to transfer phage particles from individual clear plaques into 0.5-mL microcentrifuge tubes containing all components of the original PCR reaction mixture minus the template. Vortex lightly and add mineral oil. Heat to 95°C for 2 min to lyse phage, and then perform 25 cycles of the original temperature regimen. Analyze by agarose gel electrophoresis.

References

1. Buchman, G. W., Schuster, D. M., and Rashtchian, A. (1992) Rapid and efficient cloning of PCR products using the CloneAmp system. *Focus* **14,** 41–45.

2. Crowe, J. S., Cooper, H. J., Smith, M. A., Sims, M. J., Parker, D., and Gewert, D. (1991) Improved cloning efficiency of polymerase chain reaction (PCR) products after proteinase K digestion. *Nucleic Acids Res.* **19,** 184.

3. Krowczynska, A. M. and Henderson, M. B. (1992) Efficient purification of PCR products using ultrafiltration. *BioTechniques* **13,** 286–289.

4. Liu, Z. and Schwartz, L. M. (1992) An efficient method for blunt-end ligation of PCR products. *BioTechniques* **12,** 28–30.

5. Bennett, B. L. and Molenaar, A. J. (1994) Cloning of PCR products can be inhibited by *Taq* polymerase carryover. *BioTechniques* **16,** 32.

6. Hitti, Y. S. and Bertino, A. M. (1994) Proteinase K and T4 DNA polymerase facilitate the blunt-end subcloning of PCR products. *BioTechniques* **16,** 802.

7. Clarke, J. M. (1988) Novel non-templated nucleotide addition reactions catalysed by prokaryotic and eukaryotic DNA polymerases. *Nucleic Acids Res.* **16,** 9677–9686.

8. Wang, K., Koop, B. F., and Hood, L. (1994) A simple method using T4 DNA polymerase to clone polymerase chain reaction products. *BioTechniques* **17,** 236.

9. New England Biolabs catalog (1994) Reference Appendix, pp. 180,181.

10. Kaufman, D. L. and Evans, G. A. (1990) Restriction endonuclease cleavage at the termini of PCR products. *BioTechniques* **9,** 304.

11. Ho, S. N., Pullen, J. K., Horton, R. M., Hunt, H. D., and Pease, L. R. (1990) *DNA Protein. Eng. Tech.* **2,** 50.

12. Jung, V., Pestka, S. B., and Pestka, S. (1993) Cloning of polymerase chain reaction-generated DNA containing terminal restriction endonuclease recognition sites, in *Methods in Enzymology*, vol. 218 (Wu, R., ed.), Academic, London, pp. 357–362.

13. Pfeiffer, B. H. and Zimmerman, S. B. (1983) Polymer-stimulated ligation: Enhanced blunt- or cohesive-end ligation of DNA or deoxyribonucleotides by T4 DNA ligase in polymer solutions. *Nucleic Acids Res.* **11,** 7853–7871.

23

Sequencing PCR Products Cloned into M13 Vectors

Neil Brewis

1. Introduction

The polymerase chain reaction (PCR) facilitates the rapid in vitro amplification of target DNA segments. Numerous applications have been developed to exploit the vast potential of PCR, and many of these require sequence analysis of the DNA product. This chapter describes the dideoxy chain-termination method for sequencing PCR products cloned into M13 and plasmid vectors.

The dideoxy chain-termination method *(1)* involves enzymatic elongation of an oligonucleotide primer that is annealed to a single-stranded DNA template. Single-stranded DNA of bacteriophage M13 *(2)* or phagemid vectors *(3)* give the most consistently satisfactory sequence data. However, denatured double-stranded plasmid DNA can also serve as a suitable template and can also yield several hundred nucleotides of sequence information per reaction.

The quality of the template DNA is critical. One of the main causes of noninterpretable sequencing data is the poor quality of the template DNA. Many different methods for preparing DNA have been developed. Here, two quick "mini-prep" methods for isolating "sequencing-quality" DNA are presented. One method describes the preparation of single-stranded DNA from M13 phage (or a phagemid vector), whereas the second protocol can be used to prepare double-stranded DNA from a plasmid clone.

Once the DNA template has been prepared, there are three steps in the sequencing protocol. The first involves the annealing of the oligonucleotide primer to the template DNA. If double-stranded DNA is used, the primer-annealing step must be preceded by a denaturation step using sodium hydroxide. In the second step, the primer is extended in the presence of deoxynucleotides by the action of a DNA polymerase, such as Sequenase 2.0. In the final step, the reaction mixture is divided into four separate tubes, each containing a dif-

From: *Methods in Molecular Biology, Vol. 65: PCR Sequencing Protocols*
Edited by: R. Rapley Humana Press Inc., Totowa, NJ

ferent dideoxynucleotide. The incorporation of a dideoxy nucleotide molecule terminates the growth of the DNA strand. The reactions are stopped by the addition of formamide and EDTA. Following heat denaturation, the reactions can be loaded onto a sequencing gel.

Sequenase version 2.0 is an excellent polymerase for DNA sequencing. A genetically engineered version of bacteriophage T7 DNA polymerase, Sequenase 2.0 exhibits no 3'–5' exonuclease activity, high processivity, and efficiently incorporates nucleotide analogs, such as α-thio-dATP, dideoxy nucleotides, deoxy inositol triphosphate, and deaza-derivatives of deoxynucleotides.

2. Materials

All solutions should be made from reagents of molecular biology grade in purified water.

2.1. General Reagents

1. Phenol: Use nucleic-acid-grade phenol equilibrated with 10 mM Tris-HCl, pH 8.0. Store with 0.1% (w/v) hydroxyquinoline at 4°C. Note: phenol is highly corrosive.
2. Chloroform.
3. 3M sodium acetate: adjust to pH 5.2 using acetic acid. Autoclave and store at room temperature.
4. 70 and 100% ethanol.

2.2. Preparation of Single-Stranded DNA

1. 2X YT broth: 1.6% (w/v) bacto-tryptone, 1% (w/v) bacto-yeast extract, 0.5% (w/v) NaCl. Adjust pH to 7.0 using 5M NaOH, make up to 1 L, and sterilize by autoclaving.
2. Stock solutions of ampicillin and kanamycin can both be prepared in water at concentrations of 50 and 10 mg/mL, respectively. Filter-sterilize and store at –20°C.
3. PEG/NaCl: 20% (w/v) PEG (mol wt 6000–8000), 2.5M sodium chloride. Sterilize by filtration and store at 4°C.

2.3. Preparation of Double-Stranded DNA

1. LB medium: 1% (w/v) bacto-tryptone, 1% (w/v) NaCl, 0.5% (w/v) bacto-yeast extract in water. Adjust pH to 7.0 using 5M NaOH, make up to 1 L, and sterilize by autoclaving.
2. Ampicillin stock: 50 mg/mL in water. Sterilize by filtration and store at –20°C.
3. 50 mM glucose, 25 mM Tris-HCl, pH 8.0, 10 mM EDTA. Autoclave and store at 4°C.
4. 0.2N NaOH/1% (w/v) SDS: Prepare fresh from stock solutions.
5. Potassium acetate, pH 5.2: Mix 120 mL 5M potassium acetate, 23 mL acetic acid, and 57 mL water. Store at 4°C.
6. RNase: Dissolve pancreatic RNase at 10 mg/mL in 15 mM NaCl, 10 mM Tris-HCl, pH 7.5. Heat to 100°C for 15 min, and allow to cool to room temperature. Store at –20°C.

2.4. Sequencing Reaction

1. Sequenase buffer (5X concentrate): 200 mM Tris-HCl, pH 7.5, 250 mM NaCl, 100 mM MgCl$_2$. Store at –20°C.
2. Oligonucleotide primer dissolved in 20 mM Tris-HCl, pH 8.0, 0.1 mM EDTA (*see* Note 1). Store at –20°C.
3. 1M NaOH/1 mM EDTA: Store at 4°C.
4. 2M ammonium acetate: Adjust pH to 5.4 with acetic acid. Filter-sterilize and store at room temperature.
5. Labeling mix (5X concentrate): 7.5 μM dCTP, 7.5 μM dGTP, 7.5 μM dTTP. Store at –20°C.
6. Sequenase Version 2.0 enzyme (13 U/μL) from US Biochemicals (Cleveland, OH). Store at –20°C.
7. Enzyme dilution buffer: 10 mM Tris-HCl, pH 7.5, 5 mM dithiothreitol (DTT), 0.5 mg/mL bovine serum albumin. Store at –20°C.
8. Dithiothreitol 0.1M: Sterilize by filtration and store at –20°C.
9. [^{35}S]dATPαS (10 μM, 10 μCi/μL). Store at –20°C. Radioactive material should be handled and disposed of accordingly.
10. ddATP termination mix: 80 μM each dATP, dCTP, dGTP, dTTP, 8 μM ddATP, 50 mM NaCl.
11. ddCTP termination mix: 80 μM each dATP, dCTP, dGTP, dTTP, 8 μM ddCTP, 50 mM NaCl.
12. ddGTP termination mix: 80 μM each dATP, dCTP, dGTP, dTTP, 8 μM ddGTP, 50 mM NaCl.
13. ddTTP termination mix: 80 μM each dATP, dCTP, dGTP, dTTP, 8 μM ddTTP, 50 mM NaCl.
14. Formamide stop solution: 95% (w/v) analytical-grade formamide, 20 mM EDTA, pH 8.0, 0.05% (w/v) bromophenol blue, 0.05% (w/v) xylene cyanol FF. US Biochemicals offers a sequencing kit that contains all buffers, nucleotide mixes, and Sequenase enzyme required for sequencing. Control DNA and primer are also included. Alternatively, nucleotide mixes can be purchased separately.

3. Methods

3.1. Preparation of DNA Template

Preparation of single-stranded DNA from M13 phage (*see* Note 2):

1. Prepare an overnight culture of a suitable host strain, such as JM109 (*see* Note 3) in 4 mL 2X TY.
2. Dilute the overnight culture 1 in 100 in 2X TY, and aliquot 2 mL into a sterile 15-mL culture tube.
3. Touch the surface of a single plaque with a toothpick, and shake the toothpick end in the diluted bacterial culture.
4. Incubate at 37°C for 5 h with vigorous shaking (*see* Note 4).
5. Spin a 1.5-mL aliquot in microfuge tube for 5 min.

6. Carefully remove the supernatant to a clean tube, and add 200 μL 20% PEG/2.5*M* NaCl. Vortex and leave at room temperature for at least 15 min.
7. Spin for 5 min in a microfuge, and carefully remove all the supernatant (*see* Note 5).
8. Resuspend the pellet in 100 μL TE by vortexing.
9. Add 50 μL phenol, vortex, let stand for 5 min, and mix again with further vortexing.
10. Spin for 2 min at high speed in a microfuge, and in a new tube, extract the aqueous phase with 50 μL chloroform.
11. In a new tube, add 10 μL of 3*M* sodium acetate, pH 5.2, and 250 μL ethanol.
12. Centrifuge for 5 min at high speed in a microfuge.
13. Wash the DNA pellet with 400 μL 70% ethanol, and spin for 2 min.
14. Dry the pellet with the tube open on the bench or under reduced pressure in a SpeedVac.
15. Resuspend in 30 μL TE.

Run a small aliquot on an agarose gel to check the quality and yield of DNA (*see* Note 6).

3.2. Preparation of Double-Stranded DNA

1. Prepare an overnight culture from a single transformed bacterial colony in 4 mL LB supplemented with 50 μg/mL ampicillin.
2. Decant 1.5 mL of culture into a microfuge tube, and spin for 1 min in a microfuge. Aspirate the supernatant, and in the same tube, pellet a further aliquot of the culture. Completely aspirate the supernatant.
3. Resuspend the bacterial pellet in 150 μL 50 m*M* glucose, 25 m*M* Tris-HCl, pH 8.0, 10 m*M* EDTA.
4. Add 300 μL of 0.2*M* NaOH/1% SDS, and invert several times before placing on ice for 5 min.
5. Add 225 μL 3*M* potassium acetate, pH 4.8, invert several times, and place on ice for another 5 min.
6. Spin at high speed in a microfuge for 5 min.
7. Extract the supernatant with 600 μL of 1:1 phenol:chloroform.
8. In a new tube, add 675 μL ethanol, briefly vortex, and pellet the DNA for 5 min using a microfuge at high speed.
9. Wash the DNA pellet with 400 μL 70% ethanol, and spin for 2 min.
10. Dry the pellet with the tube open on the bench or under reduced pressure in a SpeedVac.
11. Resuspend in 30 μL TE containing 20 μg/mL RNase.

Run a small aliquot on an agarose gel to check the quality and yield of DNA (see Note 7).

3.3. Annealing Reaction

Single-Stranded DNA Templates:

1. In a microfuge tube mix the following:

Single-stranded DNA template	1 μg (*see* Note 8)
Primer	0.5 pmol (*see* Note 9)

Sequenase buffer (5X)	2 µL
Water	to 10 µL

2. Incubate in a 65°C water bath for 2 min, and then allow to cool slowly to 37°C over 30 min (*see* Note 10).

3.4. Double-Stranded DNA Templates

1. In a microfuge tube mix the following:

Double-stranded DNA template	3 µg (*see* Note 11)
$1M$ NaOH/1 mM EDTA	4 µL
Water	to 20 µL

2. Incubate for 5 min at room temperature.
3. Add 2 µL $2M$ ammonium acetate, pH 5.4, mix, and add 55 µL ethanol. Place the tube on ice for 30 min.
4. Spin in a microfuge at high speed for 15 min at 4°C, and carefully remove and discard the supernatant.
5. Wash the pellet with 200 µL 70% ethanol, spin in a microfuge at high speed for 5 min, and remove and discard the supernatant. Dry the pellet with the tube open on the bench or under reduced pressure in a SpeedVac. Use immediately, or store at −70°C.
6. Dissolve the DNA in the following:

Primer	1.5 pmol (*see* Note 9)
Sequenase buffer (5X)	2 µL
Water	to 10 µL

7. Incubate at 65°C for 2 min, and then place at 37°C for 20 min.

3.5. Labeling Reaction

1. Dilute the labeling mix 1 in 5 with water (*see* Note 12).
2. Dilute the Sequenase 1 in 8 with ice-cold enzyme dilution buffer. Store on ice.
3. Add the following to 10 µL of solution of annealed primer-template:

Dithiothreitol 0.1M	1 µL
Diluted labeling mix	2 µL
[^{35}S]dATPαS (10 µCi/µL)	0.5 µL
Diluted Sequenase enzyme	2 µL

4. Mix (*see* Note 13) and incubate at room temperature for 5 min.

3.6. Termination Reactions

1. Label four tubes "A," "C," "G," and "T" (*see* Note 14).
2. Pipet 2.5 µL of the appropriate termination mix into the bottom of each tube (i.e., ddATP termination mix into the tube labeled "A," and so forth).
3. Prewarm the four tubes to 37°C in a water bath.
4. Add 3.5 µL of the labeling reaction to each of the four tubes and mix.
5. Incubate at 37°C for 5 min.
6. Add 4 µL of formamide stop solution and mix. Store on ice until ready to load the sequencing gel. Samples can be stored at −20°C for 1 wk.

7. Heat the tubes to 80°C for 3 min immediately prior to loading a 2.5-µL aliquot onto the sequencing gel.

4. Notes

1. A variety of so-called universal primers and reverse primers, complementary to sequences in the poly-linker region of M13 cloning phages and plasmids, are commercially available. Custom primers can also be made using an automatic DNA chemical synthesizer (*see* ref. *4*). These should be about 30 nucleotides away from the sequence of interest. Generally, they should be about 17–19 oligomers with a 50% G + C content. Try to avoid palindromic sequences.

2. Single-stranded DNA can also be prepared from phagemid vectors, such as pBluescript, using the following procedure. Inoculate a single colony of transformed *Escherichia coli* XL1 blue cells into 4 mL 2X YT supplemented with 75 µg/mL ampicillin and VCSM13 helper phage at about 5×10^6 PFU/mL. Grow with vigorous aeration for 1–2 h at 37°C. Add kanamycin to 70 µg/mL, and continue to incubate overnight. Continue at step 5.

3. The host bacteria should be streaked on a minimal agar plate, grown overnight, and a single colony used as the inoculum.

4. Longer than 5 h or temperatures higher than 37°C can lead to reduced yields.

5. After removing most of the supernatant, a short second spin may be advantageous to help remove all the PEG solution.

6. The yield of single-stranded DNA is about 5 µg/mL of culture.

7. The yield of double-stranded DNA is about 4 µg/mL of culture.

8. A small excess of DNA, e.g., 2 µg, can be used and may help reading sequences close to the primer.

9. 0.5 and 1.5 pmol of a 17 oligonucleotide is equivalent to approx 2.8 and 8.4 ng, respectively.

10. The required rate of cooling can be achieved by incubation in a small beaker of 65°C water that is allowed to stand at room temperature for about 30 min.

11. A small excess of DNA, e.g., 5 µg, can be used and may help reading sequences close to the primer.

12. In order to read sequences farther away from the primer, the labeling mix can be diluted less and even used undiluted.

13. Mix by gently pipeting several times up and down.

14. Multiwell microtiter plates with U-shaped wells are ideal for performing the termination reactions. The annealing and labeling reactions are conveniently performed in microfuge tubes. When using microtiter plates, take care to prevent excessive evaporation by replacing the lid.

References

1. Sanger, F., Nicklen, S., and Coulson, A. R. (1977) DNA sequencing with chain-terminating inhibitors. *Proc. Natl. Acad. Sci. USA* **74,** 5463–5467.

2. Messing, J. (1993) M13 Cloning Vehicles: Their contribution to DNA sequencing, in *DNA Sequencing Protocols* (Griffin, H. G. and Griffin, A. M., eds.), Humana, Totowa, NJ, pp. 9–22.
3. Zagursky, R. J. and Berman, M. L. (1984) Cloning vectors that yield high levels of single-stranded DNA for rapid DNA sequencing. *Gene* **27,** 183–191.
4. Gerischer, U. and Dürre, P. (1993) Sequencing using custom designed oligonucleotides, in *DNA Sequencing Protocols* (Griffin, H. G. and Griffin, A. M., eds.), Humana, Totowa, NJ, pp. 75–82.

24

Genomic Amplification
with Transcript Sequencing (GAWTS)*

Tammy Lind, Erik C. Thorland, and Steve S. Sommer

1. Introduction

Genomic amplification with transcript sequencing (GAWTS) *(1,2)* is a generally applicable method for direct sequencing of PCR material. GAWTS is centered around the attachment of a phage promoter sequence (T7, Sp6, or T3) to the 5'-end of one or both PCR primers. The phage promoter sequence allows the PCR product to be transcribed into RNA. Subsequently, the RNA is utilized as a single-stranded template for dideoxynucleotide sequencing with AMV reverse transcriptase (Fig. 1).

GAWTS has several advantages over other direct sequencing methods:

1. The transcription step provides an amplification, subsequent to PCR, of up to 1000-fold of the region of interest, eliminating further purification following the PCR.
2. The additional amplification at the transcriptional level compensates for low-yield PCR, and provides a strong sequencing signal.
3. A single-stranded template provides increased sequence reproducibility relative to a double-stranded template.

GAWTS lends itself to automation with a robotic device because cumbersome methods, such as ethanol precipitation or centrifugation, are not required. Under optimal conditions, sequence autoradiograms sometimes can be read accurately after 2 h of exposure at room temperature. In these cases, a sequence reaction that is 1% of optimal intensity may be salvaged by autoradiography

---

*The information in this chapter has been adapted from Sommer and Vielhaber (1994) *(2a)* and supplemented with knowledge gained from further experience with GAWTS.

From: *Methods in Molecular Biology, Vol. 65: PCR Sequencing Protocols*
Edited by: R. Rapley Humana Press Inc., Totowa, NJ

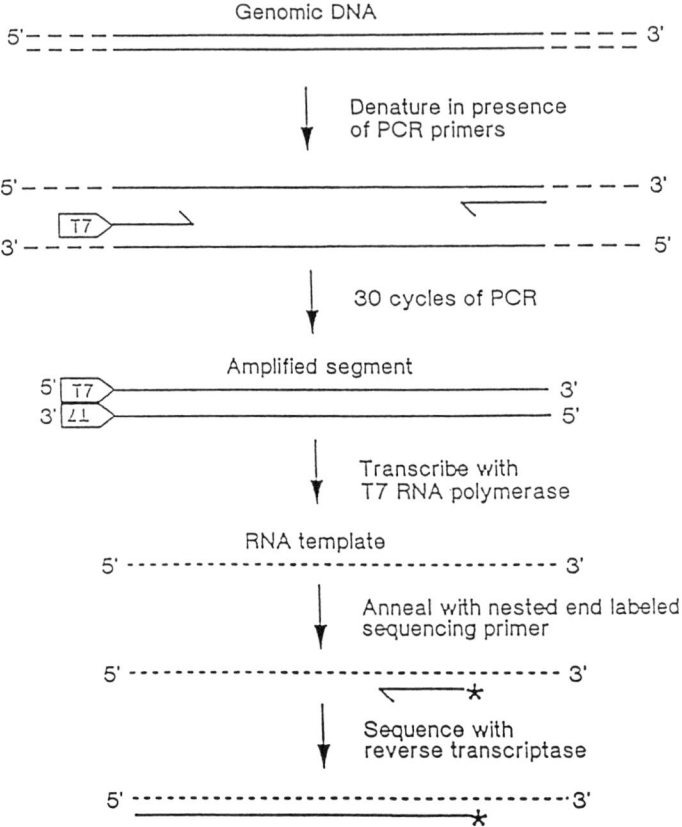

Fig. 1. Schematic of GAWTS. GAWTS consists of the following three steps: (1) PCR, in which one or both oligonucleotides contain a phage promoter in addition to a sequence targeting the primer to the region to be amplified, (2) transcription with the phage promoter, and (3) dideoxy sequencing of the transcript with reverse transcriptase that is primed with a nested (internal) oligonucleotide. Reprinted with permission of Academic Press from Sommer et al. *(2)*.

with an intensifying screen for 2 d at −70°C. These advantages make GAWTS a "forgiving" and technically robust method.

The disadvantages of GAWTS vs other methods include limited choice of sequencing enzyme and the additional expense of synthesizing oligonucleotides with 23-bp promoter sequences. Degradation of RNA is not a problem; a great abundance of template RNA is generated in an essentially nuclease-free environment. This is in contrast to mRNA, which is often unstable because it is present at low levels and is isolated from cells that contain large amounts of nucleases.

In our laboratory, GAWTS has been the method utilized for determining the causative mutation in more than 500 patients with hemophilia B. Sequence data have also been obtained for other human genes, such as p53, transthyretin, and the dopamine D_2 receptor *(3–5)*. In total, more than 1 Mb of genomic sequence has been generated with GAWTS. Sequence has also been obtained from mRNA by generating one strand of cDNA and diluting the sample in the PCR reaction *(6)*.

Other phage promoter-based methods have been developed for rapid screening for the presence of mutations and for sequencing from nongenomic DNA. RNA single-strand conformation polymorphism (rSSCP) *(7,8)* is a modification of single-strand conformation polymorphism (SSCP) *(9)*. A direct comparison between the methods indicates that rSSCP is significantly more efficient than SSCP at detecting mutations, but not as good as direct sequencing. Dideoxy fingerprinting (ddF) *(10)* is a hybrid of SSCP and GAWTS in which one lane of a dideoxy sequencing reaction is electrophoresed through a nondenaturing gel. ddF detected 100% of a group of 84 mutations with no false positives. The analysis included a blinded comparison with direct sequencing for the analysis of seven regions of likely functional significance in the factor IX gene of 30 patients with hemophilia B. Subsequent blinded analyses with the factor IX and p53 genes also detected 100% of the sequence changes *(11,12)*.

GAWTS is used to sequence DNA segments when at least some of the segment sequence is previously available. Cellular RNA can be sequenced by RNA amplification with transcript sequencing (RAWTS) *(6)*. Novel DNA sequence can be obtained on an unlimited number of isolates from a cDNA or genomic library by an extension of GAWTS known as promoter ligation and transcript sequencing (PLATS) *(13)*. With PLATS, an unlimited amount of DNA sequence can be obtained in both directions with a few generic oligonucleotides. Finally, any desired segment of a protein can be translated in vitro by attaching a Kozak translation initiation sequence 3' to the phage promoter sequence of an oligonucleotide *(14)*.

2. Materials

All solutions should be made to the standard required for molecular biology. Use molecular biology grade reagents and sterile distilled water.

2.1. PCR Reagents

1. *Taq* DNA polymerase (Boehringer Mannheim, Indianapolis, IN).
2. 10X PCR buffer: 500 m*M* KCl, 100 m*M* Tris-HCl, pH 8.3, 15 m*M* MgCl$_2$. Store at –20°C in 1-mL aliquots.
3. dNTP mix: 1.25 m*M* of each dNTP (100 m*M* stocks from Boehringer Mannheim). Prepare and store at –20°C in 1-mL aliquots.

4. 10 mM MgCl$_2$.
5. Oligonucleotide primers with phage-promoter sequences attached to the 5'-end:
    T7: 5' TAA TAC GAC TCA CTA TAG GGA GA 3'
    Sp6: 5' ATT TAG GTG ACA CTA TAG AAT AG 3'
    T3: 5' AAT TAA CCC TCA CTA AAG GGA AG 3'
6. Genomic DNA in TE buffer or ddH$_2$O, concentration approx 250 ng/μL.

2.2. Transcription Reagents

1. T7 (Promega), Sp6 (Boehringer Mannheim), or T3 RNA polymerase (Promega, Madison, WI), to match the phage-promoter sequence (*see* Notes 1 and 2).
2. 5X transcription buffer: 200 mM Tris-HCl, pH 7.5, 30 mM MgCl$_2$, 10 mM spermidine, 50 mM NaCl. Store in 1-mL aliquots at –20°C.
3. Ribonucleotide mix: 2.5 mM each rNTP (100 mM stocks from Boehringer Mannheim). Make 400-μL aliquots and store at –20°C. Unstable to freeze/thaw.
4. 100 mM DTT (Promega provides with RNA polymerase).
5. RNasin (Promega) (not essential).

2.3. End-labeling Reagents

1. T4 polynucleotide kinase (Promega).
2. End-labeling buffer: 500 mM Tris-HCl, pH 7.4, 100 mM MgCl$_2$, 50 mM DTT, 1 mM spermidine. Make 100–μL aliquots and store at –20°C.
3. γ-[^{32}P] ATP, aqueous, >5000 Ci/mmol, Amersham (Arlington Heights, IL) PB10218.
4. Oligonucleotide sequencing primer, internal to PCR primers.

2.4. Sequencing Reagents

1. Annealing buffer: 250 mM KCl, 10 mM Tris-HCl, pH 8.3. Make 1-mL aliquots and store at –20°C.
2. AMV reverse transcriptase (Boehringer Mannheim).
3. Reverse transcriptase buffer, as provided with AMV RT or as follows: 24 mM Tris-HCl, pH 8.3, 16 mM MgCl$_2$, 8 mM DTT, 0.8 mM dATP, 0.4 mM dCTP, 0.8 mM dGTP, 1.2 mM dTTP, 100 μg/mL actinomycin D. Make a 135-μL mix **without actinomycin D**. Immediately prior to performing the sequencing reaction, thaw aliquot and add 15 μL of 1 mg/mL actinomycin D.
4. Dideoxy nucleotides (Boehringer Mannheim): Make 100-μL solutions from 10 mM stocks and store at –20°C (*see* Notes 3 and 4).
    1 mM ddATP.
    1 mM ddTTP.
    1 mM ddGTP.
    0.25 mM ddCTP.
5. Stop solution: 85% formamide, 25 mM EDTA, pH 8.0, 0.1% bromophenol blue, 0.1% xylene cyanol. Amounts of dye can be decreased if desired. Make 1-mL aliquots and store at –20°C.

2.5. Electrophoresis Reagents

1. 10X TBE2 buffer (pH adjusted to 8.3 with HCl or $10M$ NaOH): $0.5M$ Tris, $0.5M$ boric acid, 10 mM EDTA. (Note that these concentrations are different from those of standard 10X TBE buffers.)
2. Gel solution (standard): $7M$ urea, 6% Long Ranger gel solution (J. T. Baker, Phillipsburg, NJ), 1.2X TBE2, 625 μL 10% APS, 65 μL TEMED (*see* Note 5). Formamide gel solution: 8% Long Ranger (J. T. Baker), $7M$ Urea, 40% formamide, 1X TBE2, 750 μL 10% APS, 75 μL TEMED.
3. Running buffer: 0.6X TBE2 for standard solution, 1X TBE2 for formamide solution.

3. Method
3.1. PCR

1. Add genomic DNA (typically 250 ng) to a 25-μL reaction containing 2.5 μL of 10X PCR buffer, 4 μL of dNTP mix, 0–5 μL of 10 mM MgCl$_2$ (for final concentration of 1.5–3.5 mM), 0.63–6.3 pmol each primer (for 0.03–0.25 μM final concentration), 0.5 U of *Taq* polymerase, and ddH$_2$O to complete the 25 μL final volume.
2. Empirically determine the MgCl$_2$ and primer concentrations for each set of primers.
3. Perform 30 cycles of PCR (denaturation: 1 min at 94°C; annealing: 2 min at 50°C; elongation: 3 min at 72°C) with the Perkin-Elmer Cetus 480 automated thermal cycler.
4. After the last cycle of PCR, perform a final 10 min of elongation at 72°C.

3.2. Transcription

1. Add 3 μL of amplified PCR product to 4 μL of 5X RNA transcription buffer, 4 μL of rNTP mix, 2 μL of 100 mM DTT, RNasin (20 U), 20 U of T7 or Sp6 RNA polymerase, and ddH$_2$O to complete the 20-μL final volume. The final mixture contains 40 mM Tris-HCl (pH 7.5), 6 mM MgCl$_2$, 2 mM spermidine, 10 mM sodium chloride, 0.5 mM of the four ribonucleoside triphosphates, RNasin (1 U/μL), 10 mM DTT, and 1 U/μL of T7 or Sp6 RNA polymerase.
2. Incubate samples for 1–3 h at 37°C.

3.3. End Labeling

1. Incubate 0.1 μg of oligonucleotide in a 13 μL reaction volume containing 1 μL of 10X end-labeling buffer, 100 μCi [γ-^{32}P]ATP (5000 Ci/mmol), and 10 U of T4 polynucleotide kinase for 30 min at 37°C.
2. Heat the reaction to 65°C for 5 min to stop the reaction, and add 7 μL of water. The final mixture contains 5 ng/μL oligonucleotide, 25 mM Tris-HCl, pH 7.4, 5 mM MgCl$_2$, 2.5 mM DTT, 0.05 mM spermidine, and 5 μCi/μL [γ-^{32}P]ATP.

3.4. Sequencing Reaction

1. Add 2 μL of the transcription reaction and 1 μL of ^{32}P end-labeled sequencing primer to 10 μL of annealing buffer.

2. Denature the samples at 80°C for 3 min and then anneal at 45°C for 30–45 min (approx 5°C below the Wallace temperature [4°C × (G + C) + 2°C × (A + T)] of the oligonucleotide).

3. Label microfuge tubes with "A," "C," "G," and "T."

4. Add the following reagents: 3.3 μL of reverse transcriptase buffer containing 1 U of AMV reverse transcriptase, 1 μL of a dideoxyribonucleoside triphosphate (concentrations listed in Section 2.4.) and, finally add 2 μL of the annealed primer/RNA template solution.

5. Incubate the sample at 55°C for 45 min.

6. Stop the reaction by adding 2.5 μL of stop buffer (*see* Note 3).

3.5. Gel Electrophoresis

1. Place the sequencing reactions in a boiling water bath for 3 min.

2. Transfer immediately to ice water and load 1.5 μL onto a 48-cm, 0.4-mm sequencing gel (6% polyacrylamide or Long Ranger).

3. Electrophorese the samples for 2 h at 50 W of constant power. For a 6% Long Ranger gel, the bromphenol blue and xylene cyanol reference dyes migrate equivalent to ssDNA of 29 and 138 bp, respectively.

4. Dry the gel for 1 h, and perform autoradiography.

4. Notes

1. Much of our experience has been with a 29-base T7 phage promoter sequence. This sequence contains the 23-base canonical T7 phage promoter sequence and an additional 6 bp (GGT ACC) 5' of the 23-base sequence, which places the promoter sequence a short distance from the end of the PCR product. However, further experiments have shown the 23-bp promoter sequence works equally well (*see* Section 2.2.).

2. Additional phage promoter sequences, such as Sp6 or T3, may be attached to the other oligonucleotide for sequencing of the opposite strand of the PCR product. For many applications, sequence of one strand is sufficient. However, faint shadow bands may occasionally interfere with the identification of mutations in which heterozygosity occurs at a particular position. Uncertainty concerning heterozygosity can be eliminated by sequencing the other strand of DNA. Performing rSSCP and ddF on both strands can help to identify mutations that may have been difficult to detect by analyzing only one strand (*see* Section 2.2.).

3. Extensive experience with GAWTS has resulted in the identification of certain problems that may lead to suboptimal results. Often these problems can be easily remedied. For example, a smearing of the sequence is observed at the top of the sequencing gel when an abundance of transcript is utilized in the sequencing reaction. Smearing can be eliminated by diluting the transcript 10-fold, or by stopping the sequencing reactions with a six- to eightfold excess of stop buffer (15–20 μL) (*see* Section 3.4.). Also, occasional shadow bands are observed on a GAWTS sequencing autoradiogram when the sequence contains "T" followed by "G". A reduction in the concentration of transcript or reverse transcriptase, or

the use of terminal deoxynucleotidyl transferase *(15)* may help to eliminate the "T shadow bands" (*see* Section 3.4.). For segments with high G + C content, a higher ratio of dNTP to ddNTP may be required in the "G" and "C" reactions (*see* Section 2.4.).

4. If one lane of a reaction is very strong at the bottom of the gel, but fades out at the top, a decrease in the concentration of that ddNTP will balance the intensity. Conversely, an increase in the concentration of a ddNTP will correct a reaction that is weak at the bottom, but too strong on top (*see* Section 2.4.).

5. For high-GC sequences, a formamide-based gel will eliminate or significantly reduce the compressions that tend to appear. These gels are more sensitive to temperature; it is important to preheat the gel to 55°C before loading. The gels must be fixed prior to drying by thorough rinsing with an aqueous solution of 10% methanol and 10% acetic acid. Allowing time for the gel to air-dry after rinsing (~15 min) seems to simplify the task of pulling the gel onto filter paper. Formamide gels will require more time for electrophoresis, and the reference dyes may migrate differently. The protocol for formamide gels is described more completely in the directions that are provided with Long Ranger gel solution (see Section 2.5.).

References

1. Stoflet, E. S., Koeberl, D. D., Sarkar, G., and Sommer, S. S. (1988) Genomic amplification with transcript sequencing. *Science* **239,** 491–494.
2. Sommer, S. S., Sarkar, G., Koeberl, D. D., Bottema, C. D. K., Buerstedde, J., Schowalter, D. B., and Cassady, J. D. (1990): Direct sequencing with the aid of phage promoters, in *PCR Protocols: A Guide to Methods and Applications* (Innis, M. A., Gelfand, D. H., Sninsky, J. J., White, T. J., eds.), Academic, NY, pp. 197–205.
2a. Sommer, S. S., and Vielhaber, E. L. (1994) Phage promoter-based methods for sequencing and screening for mutations, in *Polymerase Chain Reaction* (Mullis, K. B., Ferré F., and Gibbs, R. A., eds.), Birkhauser, Boston, pp. 214–221.
3. Sommer, S. S., Cunningham, J., McGovern, R. M., Saitoh, S., Schroeder, J. J., Wold, L. E., and Kovach, J. S. (1992) Pattern of p53 gene mutations in breast cancers of women of the Midwestern United States. *J. Natl. Cancer Inst.* **84,** 246–252.
4. Ii, S., Minnerath, S., Ii, K., Dyck, P. J., and Sommer, S. S. (1991) Two tiered DNA-based diagnosis of transthyretin amyloidosis reveals two novel point mutations. *Neurology* **41,** 893–898.
5. Sarkar, G., Kapelner, S., Grandy, D. K., Marchionni, M., Civelli, O., Sobell, J., Heston, L., and Sommer, S. S. (1991) Direct sequencing of the dopamine D2 receptor (DRD2) in schizophrenics reveals three polymorphisms but no structural change in the receptor. *Genomics* **11,** 8–14.
6. Sarkar, G. and Sommer, S. S. (1988) RNA amplification with transcript sequencing (RAWTS). *Nucleic Acids Res.* **16,** 5197.
7. Sarkar, G., Yoon, H., and Sommer, S. S. (1992) Screening for mutations by

RNA single-strand conformation polymorphism (rSSCP): comparison with DNA-SSCP. *Nucleic Acids Res.* **20,** 871–878.

8. Danenberg, P. V., Horikoshi, T., Volkenandt, M., Danenberg, K., Lenz, H., Shea, L. C. C., Dicker, A. P., Simoneau, A., Jones, P. A., and Bertino, J. R. (1992) Detection of point mutations in human DNA by analysis of RNA conformation polymorphism(s). *Nucleic Acids Res.* **20,** 573–579.

9. Orita, M., Suzuki, Y., Sekiya, T., and Hayashi, K. (1989) Rapid and sensitive detection of point mutations and DNA polymorphisms using the polymerase chain reaction. *Genomics* **5,** 874–879.

10. Sarkar, G., Yoon, H., and Sommer, S. S. (1992) Dideoxy fingerprinting (ddF): a rapid and efficient screen for the presence of mutations. *Genomics* **13,** 441–443.

11. Liu, Q. and Sommer, S. S. (1994) Parameters affecting the sensitivities of dideoxy fingerprinting and SSCP. *PCR Methods and Applications* **4,** 97–108.

12. Blaszyk, H., Hartmann, A., Schroeder, J. J., McGovern, R. M., Sommer, S. S., and Kovach, J. S. (1995) Rapid and efficient screening for p53 gene mutations by dideoxy fingerprinting (ddF). *BioTechniques* **18,** 256–260.

13. Schowalter, D. B., Toft, D. O., and Sommer, S. S. (1990) A method of sequencing without subcloning and its application to the identification of a novel ORF with a sequence suggestive of a transcriptional regulator in the water mold. *Achlya ambisexualis. Genomics* **6,** 23–32.

14. Sarkar, G. and Sommer, S. S. (1989) Access to an mRNA sequence or its protein product is not limited by tissue or species specificity. *Science* **244,** 331–334.

15. DeBorde, D. C., Naeve, C. W., Herlocher, M. L., and Maassab, H. F. (1986) Resolution of a common RNA sequencing ambiguity by terminal deoxynucleotidyl transferase. *Anal. Biochem.* **157,** 275–282.

25

DNA Rescue by the Vectorette Method

Marcia A. McAleer, Alison Coffey, and Ian Dunham

1. Introduction

A major advance in physical mapping of the human genome was the development of yeast artificial chromosome (YAC) vectors *(1)*. This has enabled the cloning of pieces of DNA several hundred kilobases in length *(2)*. The availability of such large cloned genomic DNA fragments means that by ordering a series of overlapping YAC clones, a contiguous stretch of DNA, several megabases in length, can be isolated around a genomic region of interest (e.g., the region of a chromosome linked to a particular disease gene). The successful isolation of terminal sequences of a given YAC can be very useful in assembling an ordered "contig" of YAC clones. Such terminal clones may be used directly as hybridization probes or sequenced and used to generate sequence tagged sites (STSs) to identify overlaps between, and isolate other, members of the contig. Several methods have been used to this end, including PCR with vector-specific primers in combination with primers designed either for repetitive elements, such as *Alu* sequences *(3)*, or in combination with random nonspecific primers *(4)*. However, these techniques rely on a suitable repetitive element or random primer sequence occurring close enough to the end of the YAC so as to be amplified by PCR. Furthermore, probes isolated in this manner may well contain highly repetitive sequences that, if unsuccessfully blocked, will increase nonspecific signal in any subsequent hybridization procedures *(5)*.

The vectorette method was originally described by Riley et al. *(6)*. YAC DNA is digested with a restriction enzyme, and the resulting fragments are ligated to a linker molecule to create a vectorette "library," i.e., a complex mixture of restriction fragments with linker ligated to each end. Within this library are fragments that contain the YAC vector/genomic DNA junction,

From: *Methods in Molecular Biology, Vol. 65: PCR Sequencing Protocols*
Edited by: R. Rapley Humana Press Inc., Totowa, NJ

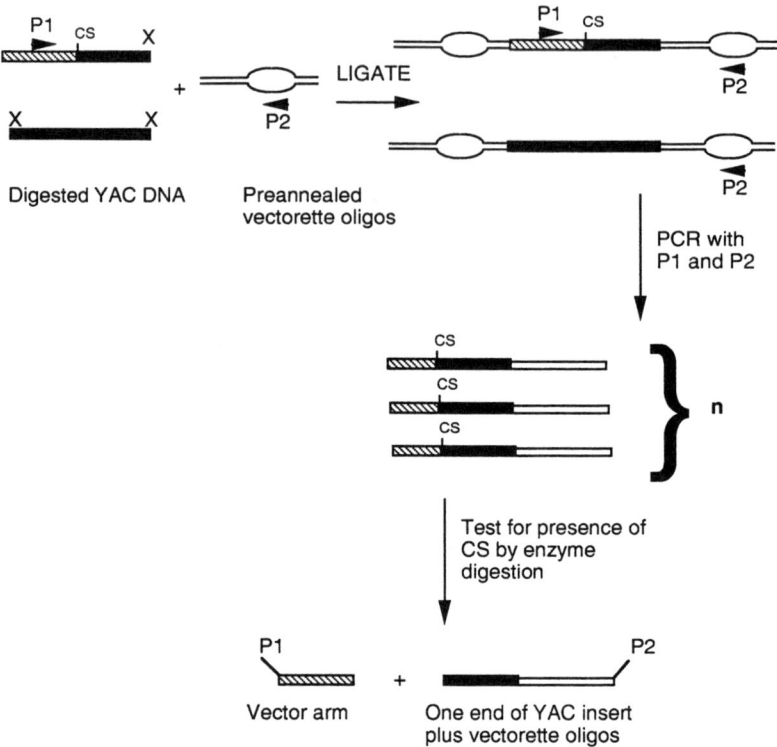

Fig. 1. A schematic representation of the vectorette method. Solid boxes represent genomic DNA, and the hatched boxes represents YAC vector sequence. YAC DNA is digested with a restriction enzyme, X. Following ligation to annealed vectorette oligos, products are amplified with a vectorette-specific primer (P2) and a primer specific for one or other of the YAC vector arms (P1). Only fragments containing vector/insert junction are amplified. Confirmation of the presence of the cloning site (CS) within the amplified fragment can be obtained by digestion of the hybrid fragment with the enzyme that cuts at the cloning site, releasing a fragment diagnostic of the vector arm (Table 2) together with one or more fragments corresponding to the genomic DNA insert. CS = cloning site.

which includes the terminal sequences of the YAC (Fig. 1). The linker molecule consists of two long (>50 nucleotides) preannealed oligonucleotides incorporating a suitable 5'-overhang corresponding to the restriction enzyme used in the initial YAC digest. Blunt-ended linkers may also be used. Although the oligonucleotides comprising the linker are complementary at the 5'- and 3'-ends, there is a region of noncomplementarity in the middle where the two strands are unable to pair and a vectorette "bubble" is formed. The PCR is then performed on this mixture using one of two vector-specific primers (designed either for the centric or acentric vector arms) in combination with a linker-specific primer. The linker-specific primer corresponds to the sequence of the linker ligated to the 5'-end of each DNA strand and has no complement on the other strand of the "bubble." It is therefore unable to anneal to template until the complementary sequence has been generated by priming off the vector-specific sequence. Thus, only those fragments containing binding sites for the vector-specific primer (i.e., DNA including and immediately adjacent to the cloning site of the YAC vector) will be successfully amplified by the PCR. The amplification products may then be used as DNA probes, for DNA sequencing, or may be cloned into a suitable vector.

A recent adaptation of the vectorette method has been used to isolate possible gene fragments from selected regions of the genome without prior knowledge of gene sequence *(7)*. This method is termed Island Rescue PCR (IRP), and relies on the fact that nearly all housekeeping genes and over 40% of tissue-specific genes have a CpG island in or near the 5'-end of the gene *(8)*. Such CpG islands have a significantly increased C + G content compared to the bulk of genomic DNA. These CpG islands can be detected in native human genomic DNA, by rare-cutting restriction endonucleases that recognize unmethylated CpG-containing sequences. The principles of the vectorette method described above are used, except the YAC DNA in this instance is digested with restriction endonucleases that specifically recognize CpG-containing sequences, e.g., *Sac*II, *Eag*I. Therefore YAC DNA will be cut at CpG-rich sites, which may be associated with a gene. The mixture is then ligated to the preannealed vectorette oligos, and PCR in this instance is driven by an *Alu*-specific primer together with the vectorette oligo described above. Northern blot analysis may then be used to test that amplified sequences are associated with expressed mRNAs. There are two main drawbacks to this method. First, since DNA in yeast is not differentially methylated, all CpG-containing restriction sites will be cut whether or not they are associated with an unmethylated island in native genomic DNA. Therefore, a portion of the amplified fragments may not be associated with an expressed mRNA. Second, as with all *Alu*-PCR based methods, there is a requirement for an *Alu* sequence close enough to the restriction site to allow amplification by the *Taq* polymerase. However, in terms of transcript mapping, where the previously described methods (direct selection/cDNA enrichment *[9]*, exon trapping *[10]*, probing cDNA libraries directly with radiolabeled YAC DNA *[11]*) all have limitations, IRP may prove to be a rapid and useful technique for the identification of transcriptional units within complex sources of DNA.

Although the vectorette method was originally developed for rescuing the vector-insert junctions of YACs, it may be used to isolate sequences adjacent to any known sequence (e.g., the identification of intron/exon boundaries in a

Table 1
Oligonucleotide Sequences for Vectorette PCR

Vectorette oligonucleotides (for blunt-ended ligations)
   "Top" strand
     CAAGGAGAGGACGCTGTCTGTCGAAGGTAAGGAACGGACGAGAGA
      AGGGAGAG
   "Bottom" strand
     CTCTCCCTTCTCGAATCGTAACCGTTCGTACGAGAATCGCTGTCCTC
      TCCTTG
Universal vectorette primer 224
   CGAATCGTAACCGTTCGTACGAGAATCGCT
pYAC4-specific primers
   Centric ("left") arm

1089	CACCCGTTCTCGGAGCACTGTCCGACCGC
Sup4-2	GTTGGTTTAAGGCGCAAGAC
pYACL	AATTTATCACTACGGAATTC

   Acentric ("right") arm

1091	ATATAGGCGCCAGCAACCGCACCTGTGGCG
Sup4-3	GTCGAACGCCCGATCTCAAG
pYACR	CCGATCTCAAGATTACGGAATTC

All oligonucleotide sequences are written in the 5'→3' direction.

specified gene *[12]*). This chapter describes in detail the application of the vectorette method to isolating terminal sequences from YACs.

2. Materials

All solutions should be made to the standard required for molecular biology, i.e., using sterile distilled water and molecular-biology-grade reagents.

1. T4 DNA ligase, 1 U/µL and 5X T4 DNA ligase buffer (0.25M Tris-HCl, pH 7.6, 50 mM MgCl$_2$, 5 mM ATP, 5 mM DTT, 25% [w/v] polyethylene glycol-8000) (Gibco BRL, Paisley, Scotland).
2. The sequences of the oligonucleotides used in this chapter are given in Table 1 and are taken from ref. *(6)*. The vector-specific primers are designed against pYAC4 (these can be replaced with appropriate vector primers or *Alu*-specific primers if performing IRP). The vectorette oligonucleotides described are suitable for blunt-ended ligations. If desired, a suitable overhang at the 5'-end of the "top" strand oligonucleotide may be incorporated to facilitate "sticky ended" ligations.

   Oligonucleotides were synthesized by phosphoramidite chemistry on an Applied Biosystems 392 DNA/RNA synthesizer. After deprotection (7 h at 55°C), oligonucleotides are dried in a centrifugal evaporator (alternatively, the standard ethanol precipitation procedure may be used). Oligonucleotides used in PCR are resuspended in H$_2$O to a concentration of 20 µM. Vectorette oligonucleotides are purified by HPLC (12% polyacrylamide gel electrophoresis may also be used). Prior to use, equimolar quantities of the "top" and "bottom" oligonucleotides are preannealed in 25 mM NaCl by heating at 65°C for 5 min and left to cool to room temperature. A working concentration of 1 µM is used in ligations. All oligonucleotides are stored at –20°C.

3. PCR is performed using a GeneAmp PCR reagent kit (Perkin Elmer, Warrington, UK) in 10 mM Tris-HCl (pH 8.3), 50 mM KCl, 1.5 mM MgCl$_2$, 0.01% (w/v) gelatin containing 200 µM of each dNTP and 1.0 µM of each primer. Amplitaq is added to a concentration of 1.25 U/50 µL reaction and Perfect Match (Stratagene, Cambridge, UK) to a concentration of 5 U/50 µL reaction, and overlaid with mineral oil (Sigma, Poole, UK). DNA amplification is performed in an Omnigene thermocycler (Hybaid, Teddington, UK).

3. Methods

1. Take half an agarose plug (approx 50–100 µL containing 1–2 µg DNA) of miniprep YAC DNA (DNA in solution may also be used; *see* Note 1) and wash as follows: 3 × 20 min in 10 mM Tris-HCl, 0.1 mM EDTA, pH 7.4 (1 mL/plug) at 50°C. 1 × 20 min in 10 mM Tris-HCl, 0.1 mM EDTA, pH 7.4 (1 mL/plug) at room temperature.
2. Preincubate plugs for 30 min at 37°C in 100 µL of the appropriate enzyme buffer (*see* manufacturer's recommendation).
3. Remove buffer, and replace with 100 µL of fresh enzyme buffer containing 20–30 U of restriction enzyme (*see* Note 2), and incubate overnight at the recommended temperature (usually 37°C). After digestion, the plug may be cut into three, and one portion electrophoresed through a 1.0% agarose mini gel alongside a similar amount of untreated YAC DNA to test for complete digestion. One slice may be stored dry at 4°C and redigested if incomplete digestion has occurred.
4. Incubate 1/3 of the agarose plug from step 3 in 1 mL of 1X ligation buffer for 1 h on ice.
5. Replace with 100 µL fresh 1X ligation buffer. To this add 10 µL of preannealed blunt-ended vectorette linker (at 1 µM: *see* Section 2., step 2) i.e., 10 pmol of linker.
6. Heat to 65°C for 15 min to melt the agarose plug, and then equilibrate at 37°C (approx 5 min).
7. When the reaction mix is equilibrated, add 1 µL of T4 DNA ligase (1 U/µL) and incubate at 37°C. After 1 h add 400 µL 10 mM Tris-HCl, 0.1 mM EDTA, pH 8.0, and mix thoroughly. The vectorette library may now be stored in aliquots at –20°C.
8. Two sets of PCR mixes need to be prepared for each vectorette library constructed. The first contains a primer, 1091, specific for the "right" arm of the YAC vector (i.e., the acentric arm encoding the URA3 gene) together with the vectorette-specific oligo (224), whereas the second contains a primer, 1089, specific for the "left" arm of the YAC vector (i.e., the centric arm, which contains the CEN4 gene) together with 224 (*see* Table 1). Each reaction is carried out in 50 µL buffer described in Section 2., step 3, including 5 µL of Perfect Match (*see* Note 3 and Fig. 2) using the following cycling conditions: 94°C for 1 min, 1 cycle, followed

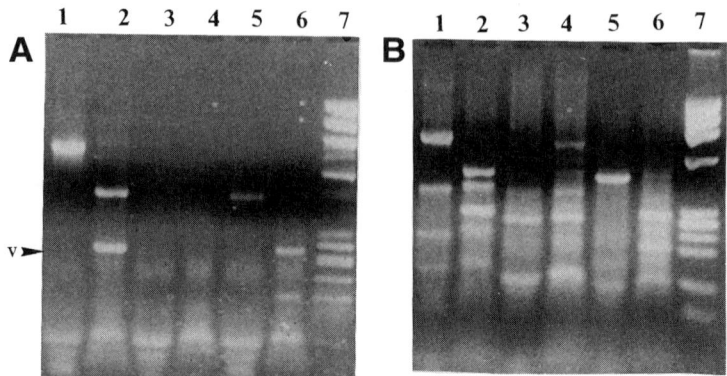

Fig. 2. Three vectorette "libraries" were created using the blunt-ended restriction enzymes: *Pvu*II (lanes 1 and 2), *Stu*I (lanes 3 and 4), and *Rsa*I (lanes 5 and 6). PCR was performed using oligos specific for the centric arm of the pYAC4, 1089, and the universal vectorette oligo, 224. In A, 5 μL of Perfect Match have been added to each PCR, whereas in B, this has been omitted. Ten microliters of untreated product were loaded on a 2.5% agarose minigel in lanes 1, 3, and 5, whereas samples in lanes 2, 4, and 6 were first digested with *Eco*RI to release the vector arm from the genomic fragment. Lane 7 contains *Hae*III fragments of ΦX RF DNA (Gibco BRL, Paisley, Scotland).

*Stu*I-digested YAC has failed to produce a PCR product (A, lanes 3 and 4), probably through the lack of an enzyme site close to the vector/insert junction. *Pvu*II- and *Rsa*I-digested YAC yields products of approx 800 and 500 bp (lanes 1 and 5), respectively, which on digestion with *Eco*RI release vector fragments (V) of the predicted size 287 bp together with the terminal *Pvu*II and *Rsa*I fragments of the YAC insert (500 and 200 bp).

by 93°C for 1 min, 65°C for 1 min, and 72°C for 3 min, 38 cycles, and followed by 72°C for 5 min, 1 cycle. For IRP, *see* Note 4 for suggested primer sequences.

9. Confirmation that PCR products originate from the terminal sequences of YAC clones can be obtained by demonstrating the presence of the YAC vector cloning site in the hybrid fragment. This is done by digesting the PCR product with a restriction enzyme that cleaves within the cloning site. To 9 μL of PCR product add 1 μL 10X restriction enzyme buffer (*see* manufacturer's recommendation), 10 U of enzyme, and incubate for 1 h at 37°C. When the vector is pYAC4, 10 U of *Eco*RI may be added directly to 9 μL of PCR product, without addition of enzyme buffer. Restriction fragments can be visualized on a 2.5% agarose minigel containing ethidium bromide (0.5 μg/mL) (ethidium bromide is a powerful mutagen and gloves should be worn at all times) (Fig. 2). The distances from the primer sequences described in Section 2., step 3 to the *Eco*RI cloning site of pYAC4 are given in Table 2.

10. A second PCR may be performed to reduce the amount of vector DNA contained in the amplified product. A nested vector-specific primer that anneals closer to the cloning site (Tables 1 and 2) is used in combination with the

Table 2
Positions of Primer Sequences Described in Table 1
with Respect to the *Eco*RI Sequence in the Cloning Site of pYAC4

Centric arm		Acentric arm	
1089→*Eco*RI	287 bp	1091→*Eco*RI	172 bp
Sup4-2→*Eco*RI	40 bp	Sup4-3→*Eco*RI	29 bp
pYACL→*Eco*RI	17 bp	pYACR→*Eco*RI	20 bp

vectorette-specific oligo. Either use 1 µL of the primary PCR or toothpick the fragment found to cut with *Eco*RI in step 9 (**not** the restriction digestion product) directly from the agarose gel into a PCR containing: for "left" arm products: Sup4-2 + 224 or pYACL + 224, and for "right" arm products: Sup4-3 + 224 or pYACR + 224. The same cycling conditions as those described in step 8 are used, but the annealing temperature is reduced to 59°C and only 20 cycles are performed. Ten microliters may be visualized on a 2.5% agarose minigel.

11. PCR products may now either be sequenced directly, radiolabeled and used as a hybridization probe (*see* Note 5), or subcloned using a suitable cloning system, such as pCR-Script™ SK(+) (Stratagene) or TA-cloning™ system (Invitrogen, Leek, The Netherlands).

4. Notes

1. Use approx 1 µg of solution DNA for each restriction enzyme digest. Reactions should be performed in the buffers recommended by the manufacturers for 4 h at the specified temperature. Before ligation (Section 3., step 5), enzymes should be heat-inactivated (65°C for 15 min is usually sufficient), extracted with phenol:chloroform (equal volume of ratio 1:1), ethanol-precipitated by standard methods (2 vol 95% ethanol with ¹/₁₀ vol 3*M* sodium acetate, pH 5.6) and resuspended to a concentration of 250 ng/µL. Ligations can be performed in a volume of 10 µL with 1 µL preannealed vectorette oligos.

2. It is important to check that there are no recognition sites for a given restriction endonuclease between the sequences corresponding to the vector-specific primers and the cloning site. If such a site were present, it would be cleaved in the initial digest and a vector-only fragment would be amplified. Suitable enzymes for pYAC4 are *Rsa*I, *Pvu*II, and *Stu*I.

3. The addition of Perfect Match to the PCR reduces the number of nonspecific bands generated (compare Fig. 2A with B), although some laboratories have found little difference on its omission.

4. IRP is a variant of *Alu*-vectorette PCR that can be used to generate probes from YACs as an alternative to *Alu*-PCR. For IRP, the universal vectorette primer 224 is used together with primer sequences that recognize a human *Alu* repeat. For example: 5'-GGATTACAGGCGTGAGCCAC-3' and 5'-GATCGCGCCACTGCAC TCC-3' (both sequences taken from ref. 7). The thermocycling conditions described in Section 3., step 8 may also be used for these two sets of primers.

5. Probes generated by this method may contain highly repetitive sequences. There-
fore, it is advisable to pre-reassociate the labeled probe with total human genomic
DNA prior to any hybridization procedure. Make probe up to 250 µL with H_2O.
Add 125 µL 10 mg/mL sonicated total human DNA (Sigma) and boil for 5 min.
Snap-chill on ice for 5 min, and then add probe to hybridization as normal.

References

1. Burke, D. T., Carle, G. F., and Olson, M. V. (1987) Cloning of large segments of
exogenous DNA into yeast by means of artificial chromosome vectors. *Science*
236, 806–812.
2. Cohen, D., Chumakov, I., and Weissenbach, J. (1993) A first-generation physical
map of the human genome. *Nature* **366,** 698–701.
3. Nelson, D. L., Ledbetter, S. A., Corbo, L., Victoria, M. F., Ramirez-Soli, R.,
Webster, T. D., Ledbetter, D. H., and Caskey, C. T. (1989) *Alu* polymerase chain
reaction: A method for rapid isolation of human-specific sequences from complex
DNA sources. *Proc. Natl. Acad. Sci. USA* **86,** 6686–6690.
4. Wesley, C. S., Myers, M. P., and Young, M. W. (1994) Rapid sequential walking
from termini of cosmid, P1 and YAC inserts. *Nucleic Acids Res.* **22,** 538–539.
5. Cole, C. G., Patel, K, Shipley, J., Sheer, D., Bobrow, M., Bentley, D. R., and
Dunham, I. (1992) Identification of region-specific yeast artificial chromosomes
using pools of *Alu* element-mediated polymerase chain reaction probes labelled
via linear amplification. *Genomics* **14,** 931–938.
6. Riley, J., Ogilvie, D., Finniear, R., Jenner, D., Powell, S., Anand, R., Smith, J. C.,
and Markham, A. F. (1990) A novel, rapid method for the isolation of terminal
sequences from yeast artificial chromosome (YAC) clones. *Nucleic Acids Res.*
18, 2887–2890.
7. Valdes, J. M., Tagle, D. A., and Collins, F. S. (1994) Island rescue PCR: A rapid
and efficient method for isolating transcribed sequences from yeast artificial chromosomes and cosmids. *Proc. Natl. Acad. Sci. USA* **91,** 5377–5381.
8. Larsen, F., Gundersen, G., Lopez, R., and Prydz, H. (1992) CpG islands as gene
markers in the human genome. *Genomics* **13,** 1095–1107.
9. Lovett, M., Kere, J., and Hinton, L. (1991) Direct selection: A method for isolation of cDNAs encoded by large genomic regions. *Proc. Natl. Acad. Sci. USA* **88,**
9628–9632.
10. Buckler, A. J., Chang, D. D., Graw, S. L., Brook, J. D., Haber, D. A., Sharp, P. A.,
and Housman, D. E. (1991) Exon amplification: A strategy to isolate mammalian
genes based on RNA splicing. *Proc. Natl. Acad. Sci. USA* **88,** 4005–4009.
11. Elvin, P., Slynn, G., Black, D., Graham, A., Butler, R., Riley, J., Anand, R., and
Markham, A. F. (1990) Isolation of cDNA clones using yeast artificial chromosome probes. *Nucleic Acids Res.* **18,** 3913–3917.
12. Roberts, R. G., Coffey, A. J., Bobrow, M., and Bentley, D. R. (1993) Exon structure of the human dystrophin gene. *Genomics* **16,** 536–538.

26

Sequencing of (dA:dT) Cloned Mixed
PCR Products from Microbial Populations

Barbara Anne Hales and Craig Winstanley

1. Introduction

Because only between 1 and 10% of bacteria present in soil and aquatic environments are culturable using currently available methods, attempts to identify and quantify the nonculturable, majority population must circumvent the need for culture. Molecular biological techniques, particularly using 16S rRNA sequences, have substantial advantages over traditional culture-based methods for the characterization of natural microbial populations. One particularly interesting strategy involves the use of the polymerase chain reaction (PCR) to amplify target genes from DNA extracted directly from environmental samples. PCR-amplified target sequences can then be cloned to obtain a representative clone bank that reflects the diversity of a target sequence in the original population. After sequencing individual clones, comparative data analysis can be used to assess diversity among a natural population of microorganisms without the need to culture the organisms first.

This approach has been applied to marine *(1)*, terrestrial *(2)*, and thermophilic *(3)* environments, and has confirmed that the prokaryotic species so far cultured constitute only a very small fraction of the actual microbial population in natural environments. Recent evidence suggests that members of the Archaea are much more abundant than had been anticipated because of the limitations of previous methods for analyzing natural ecosystems *(4)*. Molecular approaches can be particularly useful when applied to the study of strictly anaerobic and other extreme environments where PCR using archaebacterial-specific 16S rRNA oligonucleotide primers can be used to study natural communities present in such environments. In this chapter we outline an approach taken to the study of microbial populations in soil. The methods described have

From: *Methods in Molecular Biology, Vol. 65: PCR Sequencing Protocols*
Edited by: R. Rapley Humana Press Inc., Totowa, NJ

been used to analyze diversity among natural methanogenic communities *(5)*, but are much more generally applicable to the study of microbial populations in soil.

In this chapter we describe the methodology for extracting bacterial DNA from soil without the need first to separate cells from soil particles. We then outline the methodology for PCR amplification of specific target sequences contained within the bacterial DNA obtained. The resulting mixture of specific target sequences must be resolved before the diversity can be studied. This is achieved by cloning the amplified sequences into a plasmid vector to produce a bank of clones that can be screened to confirm the presence of an inserted fragment of the correct size. Finally, we describe how these individual clones can be analyzed by double-stranded DNA sequencing using DNA extracted using the QIAGEN Ltd. (Surrey, UK) Plasmid Midi Kit and the US Biochemicals (Cleveland, OH) Sequenase Kit.

2. Materials

2.1. Extraction of Bacterial DNA from Soil

1. Extraction buffer: 0.1*M* phosphate buffer, pH 8.0, containing 1% sodium dodecyl sulfate (SDS).
2. TE: 10 m*M* Tris-HCl, 1 m*M* EDTA, pH 8.0.

2.2. Cloning the PCR-Amplified Products

1. SOC medium: 2% bacto-tryptone, 0.5% bacto-yeast extract, 10 m*M* NaCl, 2.5 m*M* KCl, 10 m*M* $MgCl_2$, 10 m*M* $MgSO_4$, 20 m*M* glucose.
2. Luria agar: 1% bacto-tryptone, 0.5% bacto-yeast extract, 0.5% NaCl.

2.3. Screening of Clones

1. STET buffer: 8% sucrose, 5% triton X-100, 50 m*M* Tris-HCl, pH 8.0, 50 m*M* EDTA, pH 8.0.
2. Lysozyme: 10 mg/mL (made in STET buffer).
3. Qiagen plasmid preparation solutions:
   a. P1: 50 m*M* Tris-HCl, 10 m*M* EDTA, pH 8.0, containing 0.1 mg/mL RNase.
   b. P2: 200 m*M* NaOH, 1% SDS.
   c. P3: 2.55*M* potassium acetate, pH 4.8.
   d. QBT buffer: 750 m*M* NaCl, 50 m*M* MOPS (3-N-morpholino-propanesulfonic acid, pK_a 7.2), 15% ethanol, pH 7.0, 0.15% Triton X-100.
   e. QC buffer: 1.0*M* NaCl, 50 m*M* MOPS, 15% ethanol, pH 7.0.
   f. QF buffer: 1.25*M* NaCl, 50 m*M* MOPS, 15% ethanol, pH 8.2.
4. Kanamycin (50 µg/mL made in water), ampicillin (50 µg/mL made in water), X-gal (40 µg/mL made in dimethylformamide).

2.4. Sequencing of Cloned PCR-Amplified Products

1. 5X concentrate reaction buffer: 200 m*M* Tris-HCl, pH 7.5, 100 m*M* $MgCl_2$, 250 m*M* NaCl.

2. Termination mixes: ddATP, ddTTP, ddCTP, ddGTP; termination mix ddATP contains 80 μM dGTP, 80 μM dATP, 80 μM dCTP, 80 μM dTTP, 8 μM ddATP, 50 mM NaCl.
3. 5X concentrate labeling mix: 7.5 μM dGTP, 7.5 μM dCTP, 7.5 μM dTTP.
4. Enzyme dilution buffer: 10 mM Tris-HCl, pH 7.5, 5 mM dithiothreitol, 0.5 mg/mL BSA.
5. Stop solution: 95% formamide, 20 mM EDTA, 0.05% bromophenol blue, 0.05% xylene cyanol.

3. Methods

3.1. Extraction of Bacterial DNA from Soil

Bacterial DNA can be extracted from soil using a variety of methods. The following is a slightly modified version of the method of Selenska and Klingmüller *(6)*:

1. Suspend 2 g of soil in 5 mL of extraction buffer.
2. Shake the suspension in a water bath at 70°C for 60 min. This should release cells from soil particles and cause lysis releasing DNA.
3. Centrifuge the sample at 2800g for 15 min at 4°C.
4. Remove the supernatant, keeping it on ice. Resuspend the pellet in 5 mL of extraction buffer, and shake this suspension in a water bath at 70°C for 20 min.
5. Repeat the centrifugation step (step 3) for the second suspension, and pool the supernatants. Again resuspend the pellet in 5 mL of extraction buffer, and shake this suspension in a water bath at 70°C for 20 min. Centrifuge again (as step 3), and add the supernatant to the other pooled supernatants. The three pooled supernatants should contain the optimum amount of DNA that can be recovered by this method.
6. Centrifuge the pooled supernatants at 8000g for 30 min at 4°C (*see* Note 1).
7. Transfer the supernatant to a fresh tube and add 0.1X vol of 5M NaCl followed by an equal volume of 30% polyethylene glycol 6000 to precipitate the DNA. Mix by inverting the tube, and store at 4°C overnight.
8. Pellet the DNA by centrifugation at 5000g for 40 min at 4°C.
9. Resuspend the pellet in a suitable volume (approx 8 mL) of TE, and subject to cesium chloride density gradient centrifugation as described by Sambrook et al. *(7)*.
10. Remove the DNA band, and extract with cesium chloride-saturated propan-2-ol to remove ethidium bromide. Dialyze DNA against TE (2-L volumes, changing regularly) to remove cesium chloride and subject 5 µL to agarose gel electrophoresis in order to visualize the DNA (*see* Note 2).

3.2. PCR Amplification of Target Sequence

Target DNA sequences used to assess genetic diversity are numerous and varied, including genes encoding enzymes of catabolic pathways or virulence factors, genes encoding cell-surface proteins, or 16S rRNA and 23S rRNA spacer regions. The most commonly used target DNA sequences in Prokarya and Archaea are the 16S rRNA genes, which are present in all bacteria. 16S rRNA gene sequences contain conserved and variable regions, and are widely used as a measure of phylogenetic relationships between organisms. By designing oligonucleotide primers to conserved regions of 16S rRNA genes it is possible to amplify target sequences from bacterial DNA in a mixed DNA sample. Individual amplified products can then be cloned and sequenced using 16S rRNA-variable region sequences to gain an indication of genetic diversity among the target population.

A typical PCR reaction will include (*see* Note 3):

1. 20 µL DNA sample (*see* Note 4);
2. 10 µL 10X *Taq* polymerase reaction buffer;
3. 1 µL of each oligonucleotide primer (at final concentration of 20 pmol/µL each);
4. 0.5 µL of *Taq* polymerase (5 U/µL); and
5. 67 µL sterile distilled water overlaid with mineral oil.

Following amplification, 10–20 µL of the mixture should be subjected to agarose gel electrophoresis in order to visualize the amplified product.

A typical PCR program will consist of: 94°C, 40 s (denaturation), 55°C, 1 min (annealing; *see* Note 5), 72°C, 2 min (extension), for between 25 and 40 cycles with a final extension at 72°C for 3.5–10 min. Samples should then be held at 4°C.

3.3. Cloning the PCR-Amplified Products

Amplification of target DNA sequences from environmental samples gives heterogeneous PCR products. In order to identify, analyze, or compare individual target sequences, amplified products must first be cloned. The technology available for the cloning of PCR-amplified products often relies on the fact that a single overhanging 3'dA is generated as a result of the amplification procedure with *Taq* polymerase. The TA Cloning™ kit (Invitrogen Corporation, British Bio-technology Products Ltd., Oxon, UK) includes a vector designed to contain an overhanging 3'dT that is complementary to the overhanging dA present on the amplified products. This allows direct ligation of the PCR product to the vector. Other enzymes used for amplification may not produce the overhanging 3'dA. If this is the case, amplified products must first be pretreated with *Taq* polymerase before they can be cloned by this method.

3.3.1. Ligation

The amount of target DNA to include in a ligation should lead to between a 1:1 and 1:3 molar ratio of vector:PCR product. PCR product concentrations can be estimated by comparing the product on agarose gels with standards of known concentrations. The amount of PCR product to be ligated is determined by the equation:

$$x \text{ ng PCR product} = \frac{(y \text{ bp PCR product}) (50 \text{ ng PCR vector})}{\text{size in bp of PCR vector}} \qquad (1)$$

where bp = base pairs.

This gives a 1:1 molar ratio. The ligation reaction includes:

1. 5 µL sterile distilled water.
2. 1 µL 10X ligation buffer.
3. 2 µL PCR vector (25 ng/µL).
4. 1 µL PCR product at required concentration.
5. 1 µL T4 DNA ligase.

The reaction should be incubated at 12°C for between 4 h and overnight.

3.3.2. Transformation

Ligated DNA is used to transform competent cells of *E. coli* INVαF' (TA cloning kit) by the following method:

1. Add 2 µL of 0.5*M* β-mercaptoethanol to 50 µL of competent cells, and mix gently.
2. Add 1 µL of ligation reaction, and mix gently.
3. Incubate on ice for 30 min.
4. Heat-shock by placing tube in a water bath at 42°C for 30 s.
5. Place tube on ice for 2 min.
6. Add 450 µL of SOC medium and shake vigorously (225 rpm) at 37°C for 1 h.
7. Spread 25-, 50-, and 100-µL samples onto Luria agar plates containing kanamycin (50 µg/mL) or ampicillin (50 µg/mL) to select for antibiotic resistance encoded by the plasmid vector, and X-gal (5-bromo-4-chloro-3-indolyl-β-D-galactoside) (40 µg/mL).
8. Allow plates to dry, and incubate at 37°C for 18–40 h. Colonies with no insert in the plasmid vector will produce a blue coloration as X-gal is converted to a colored product. Colonies with plasmid vector containing an insert will be white or sometimes pale blue if the insert is <500 bp in size (*see* Note 6).

3.4. Screening of Clones (see Note 7)

White colonies should be purified by streak plating. The presence of an insert band of the correct size should be confirmed by plasmid mini preparation. A number of methods are available for this. The following is based on the rapid boiling method of Holmes and Quigley *(8)*:

1. Harvest 3 mL of overnight nutrient broth culture.
2. Resuspend pellet in 0.5 mL STET buffer.
3. Add 50 µL of lysozyme, and mix by inverting the tube.
4. Place tube in a boiling water bath for 40 s.
5. Centrifuge at high speed in a microfuge for 20 min.
6. Remove supernatant (containing DNA) to a fresh tube, and add an equal volume of precooled propan-2-ol.
7. Place at −20°C for 10 min.

8. Centrifuge for 10 min (as in step 5) to pellet DNA.
9. Pour off the supernatant, and wash the pellet by adding 0.5 mL diethyl ether and microfuge at high speed for 10 min.
10. Pour off the diethyl ether, and dry pellet by placing tube with lid open in a heating block at 37°C for 5 min.
11. Resuspend pellet in 100 μL sterile distilled water.
12. Digest 3–5 μL with *Eco*RI, and subject to agarose gel electrophoresis alongside a sample of vector DNA and a size marker (*see* Note 8).

3.5. Sequencing of Cloned PCR-Amplified Products

By sequencing a number of different clones produced by the methods already described, it is possible to gain information about the genetic diversity of the original target population from which the DNA was obtained.

3.5.1. Preparation of DNA for Double-Stranded Sequencing

Purified DNA for sequencing can be obtained by isolating plasmid DNA from clones by using the QIAGEN Plasmid Midi Kit (Qiagen Ltd., Surrey, UK).

1. Grow clone overnight at 37°C in 150-mL nutrient broth, and pellet cells by centrifugation.
2. Resuspend pellet in 4 mL of buffer P1.
3. Add 4 mL of buffer P2, mix, and incubate at room temperature for 5 min.
4. Add 4 mL of buffer P3, mix gently, and centrifuge at 30,000g for 30 min at 4°C.
5. Remove supernatant and recentrifuge for 10 min as in step 4 to assure that all cell debris is removed (*see* Note 9).
6. Equilibrate Qiagen tip 100 with 3 mL of QBT buffer, and allow to drain.
7. Apply supernatant from step 5 to the tip 100 and allow the sample to pass through. DNA will bind to the column.
8. Wash the column twice with 5 mL of buffer QC.
9. DNA can be eluted from the column with 5 mL of buffer QF. Collect the eluate, and add 0.7 vol of propan-2-ol to it.
10. Recover DNA precipitate by centrifugation at 15,000g for 30 min at 4°C.
11. Discard the supernatant, and wash the pellet with 5 mL cold 70% ethanol. Air-dry for 5 min and redissolve in 140 μL of sterile distilled Hypersolv (BDH) water. Divide into 16-μL aliquots. These can be stored at –20°C for long periods until required for sequencing.
12. Denature DNA by adding 4 μL of 1*M* NaOH (made in Hypersolv water) to a 16-μL aliquot of DNA, leaving at room temperature for 5 min.
13. Add 2 μL of 3*M* potassium acetate, pH 4.6 (made in Hypersolv water), mix, and add 50 μL of absolute ethanol (precooled at –20°C).
14. Allow precipitation on ice for 15 min.
15. Pellet DNA at high speed in a microfuge for 15 min.
16. Wash pellet with 1 mL of cold 70% ethanol and centrifuge again for 10 min.
17. Dry pellet by placing tube with lid off in a heating-block at 37°C for 15 min.

18. Resuspend dried pellet in 7 µL of sterile Hypersolv water. This sample is then used directly in the sequencing reactions.

3.5.2. Sequencing of Clones

Double-stranded DNA sequencing can be carried out on DNA prepared as described in Section 3.5.1. by using the Sequenase Kit (United States Biochemical) as follows:

1. Add 2 µL of reaction buffer (5X concentrate) and 1 µL of sequencing primer (0.5–1.0 pmol; either vector primers or primers used previously for PCR) to 7 µL of DNA.
2. Anneal the primer to the target DNA by heating the mixture in a water bath or beaker to 65°C for 2 min, and allowing it to cool slowly to 30°C over 30 min. Place the mixture on ice in preparation for carrying out sequencing reactions.
3. In individual labeled tubes, place 2.5 µL of each termination mix. Place the tubes at 37°C.
4. Dilute the labeling mix 1:5 in sterile distilled Hypersolv water, and dilute the Sequenase polymerase enzyme 1:8 in the enzyme dilution buffer.
5. For the labeling reaction, add 10 µL of cold annealed DNA, 1 µL 0.1M dithiothreitol, 2 µL diluted labeling mix, 0.5 µL (^{35}S)dATP, and 2 µL diluted Sequenase polymerase to a tube.
6. Mix and incubate for 2 min at room temperature.
7. Add 3.5 µL of labeling reaction to each of the prewarmed tubes containing termination mix (step 3).
8. Mix and incubate for 3 min at 37°C.
9. Add 4 µL of stop solution to each tube to end the reaction.

These samples can be stored at –20°C until being loaded (3 µL) onto 6% wedged polyacrylamide sequencing gels. Prior to loading, samples should be heated to 75°C for at least 2 min to denature the DNA (*see* Notes 10 and 11).

4. Notes

1. Supernatants obtained will be brown in color owing to the presence of humic acids. The pellet obtained after PEG6000 precipitation may be particulate and difficult to resuspend. It is often not possible to achieve complete resuspension.
2. To check that the fluorescence under UV light is the result of DNA and not humic matter, the sample can be digested overnight at 37°C with a restriction enzyme or with DNaseI (use 20 U).
3. If amplification is unsuccessful or leads to multiple unexpected fragments, the concentration of $MgCl_2$ in the polymerase buffer and the annealing temperature should be varied to optimize conditions.
4. A dilution series of DNA (as isolated in Section 3.1.) should be used to carry out PCR amplification from a range of target DNA concentrations (usually neat to 10^{-3}× dilution; Fig. 1). Target DNA can also be provided from bacterial cultures using 1 µL of overnight nutrient broth culture made up to 20 µL with sterile distilled water.

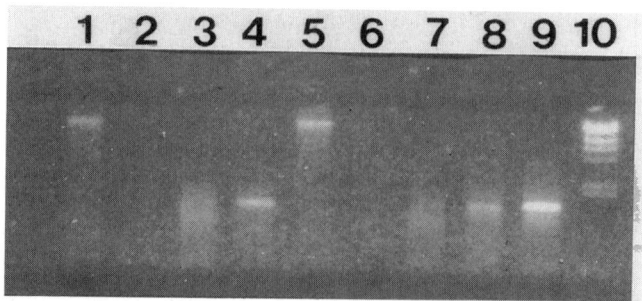

Fig. 1. PCR amplification of target sequences using DNA extracted directly from soil. The photograph shows an agarose gel of samples taken from PCR reactions carried out on DNA extracted directly from soil. Results from two different soil extracts are presented. Lanes 1–4 are from PCR reactions carried out with undiluted (lane 1), 10^{-1} diluted (lane 2), 10^{-2} diluted (lane 3), and 10^{-3} diluted (lane 4) DNA from the first extract. Lanes 5–8 are from PCR reactions carried out with undiluted (lane 5), 10^{-1} diluted (lane 6), 10^{-2} diluted (lane 7), and 10^{-3} diluted (lane 8) DNA from a second extract. Lane 9 shows a positive control consisting of an amplified product obtained by PCR amplification using a cultured organism. Lane 10 contains λ DNA digested with *Hind*III.

5. The annealing temperature will vary depending on the melting temperature (T_m) of the oligonucleotide primers used. Experiments to determine the best annealing temperature for a particular primer set should be carried out.
6. If no clones are obtained, it may be necessary first to clean the PCR-amplified DNA sample using a QIAquick PCR purification kit (Qiagen Ltd.). Increase the amount of insert DNA relative to the vector, and check the competence of cells using a routine plasmid vector, such as pUC18/19. Check that the ligation was successful by agarose gel electrophoresis of the ligation mixture prior to transformation.
7. It may also be possible to screen clones using PCR amplification performed on nutrient broth cultures (as outlined in Section 3.2.) to confirm the presence of the inserted target sequence. Care must be taken that any oligonucleotide primers used would not amplify DNA from the *E. coli* host as may be the case with 16S rRNA-targeted primers.
8. Genuine clones should give one fragment the size of the vector and additional fragments corresponding to the inserted target sequence. If the inserted sequence contains no cutting site for *Eco*RI, only two bands in all (one for vector and one for insert) will be seen. If the vector used was not from the TA Cloning Kit, other restriction enzymes may be needed to cut out the insert band.
9. In the Qiagen plasmid midi-kit extraction method, it may be possible to omit the second (10 min) centrifuge step (step 5) if the supernatant obtained in step 4 is sufficiently clear. However, any cell debris that remains at this stage may lead to blockage of the columns later.
10. When double-stranded DNA sequencing of GC-rich sequences is carried out secondary structure may lead to bands appearing in more than one lane (Fig. 2). This can be overcome by adding 0.5 μL of single-stranded DNA binding protein (SSB;

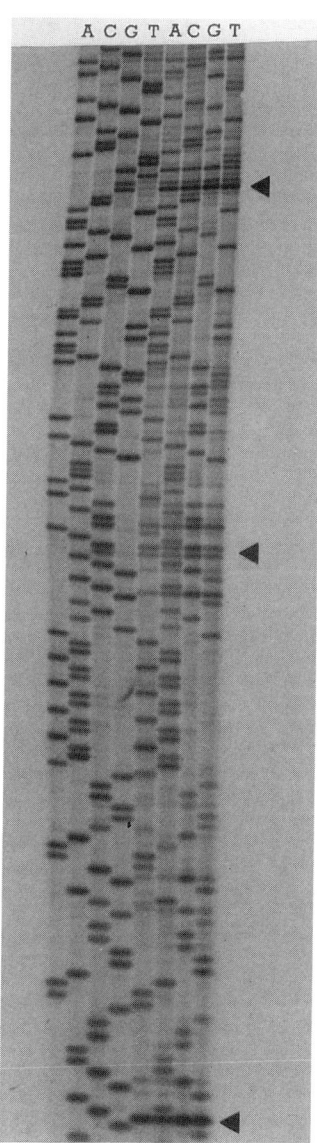

Fig. 2. Double-stranded DNA sequencing of GC-rich sequences. The sequencing gel shows the same set of reactions with (the first set of four lanes marked ACGT) and without (the second set of four lanes marked ACGT) the addition of SSB (*see* Note 8). The arrowheads indicate regions where resolution of secondary structure in the presence of SSB is particularly apparent.

1 mg/mL) to the labeling reaction. When using SSB, it is necessary to inactivate the protein prior to running the gel. This can be done by using a 4:5 mixture of stop solution:proteinase K (100 μg/mL). Five microliters of this new "stop" solution should be used to stop the sequencing reactions. The reaction times must be adhered to as strictly as possible to avoid secondary structure formation. The incubation times are the minimum recommended by the manufacturers of Sequenase (United States Biochemical). These times can be increased if secondary structure is not a problem, although the labeling step should not be longer than 5 min. Termination reactions can be run up to 50°C. This may improve results for some templates.

11. If smearing of DNA bands occurs, then the quality of DNA must be improved. The DNA preparation should be repeated.

References

1. Giovannoni, S. J., Britschgi, T., Moyer, C. L., and Field, K. G. (1990) Genetic diversity in Sargasso sea bacterioplankton. *Nature (Lond.)* **345,** 60–63.
2. Liesack, W. and Stackebrandt, E. (1992) Occurrence of novel groups of the domain *Bacteria* as revealed by analysis of genetic material isolated from an Australian terrestrial environment. *J. Bacteriol.* **174,** 5072–5078.
3. Ward, D. M., Weller, R., and Bateson, M. M. (1990) 16S rRNA sequences reveal numerous uncultured microorganisms in a natural community. *Nature (Lond.)* **345,** 63–65.
4. Olsen, G. J. (1994) Archaea, Archaea, everywhere. *Nature (Lond.)* **371,** 657.
5. Hales, B. A., Edwards, C., Ritchie, D. A., Hall, G., Pickup, R. W., and Saunders, J. R. (1996) Isolation and identification of methanogen-specific DNA from blanket bog peat using PCR amplification and sequence analysis. *Appl. Environ. Microbiol.* 62, 668–675.
6. Selenska, S. and Klingmüller, W. (1991) DNA recovery and direct detection of Tn5 sequences from soil. *Lett. Appl. Microbiol.* **13,** 21–24.
7. Sambrook, J., Fritsch, E. F., and Maniatis, T. (1989) *Molecular Cloning: a Laboratory Manual,* 2nd ed. Cold Spring Harbor Laboratory, Cold Spring Harbor, NY.
8. Holmes, D. S. and Quigley, M. (1981) A rapid boiling method for the preparation of bacterial plasmids. *Anal. Biochem.* **114,** 193–197.

Index